Erwin Dee Kord (Ed.)

Drogosławiec

Erwin Dee Kord (Ed.)

Drogosławiec

Village, Gmina Szamocin, Gmina, Chodzież County, Greater Poland Voivodeship

Solv

Imprint

Publisher:
Solv is a trademark of
International Book Market Service Ltd., 17 Rue Meldrum, Beau Bassin, 1713-01 Mauritius
Email: info@bookmarketservice.com
Website: www.bookmarketservice.com

Published in 2011

Printed in: U.S.A., U.K., Germany. This book was not produced in Mauritius.

ISBN: 978-613-8-94933-6

Contents

Articles

References

Drogosławiec

Drogosławiec	
— Village —	
Country	Poland
Voivodeship	Greater Poland
County	Poznań County
Gmina	Stęszew

Drogosławiec [drɔɡɔˈswavjɛt͡s] is a village in the administrative district of Gmina Stęszew, within Poznań County, Greater Poland Voivodeship, in west-central Poland.[1]

References

[1] "Central Statistical Office (GUS) - TERYT (National Register of Territorial Land Apportionment Journal)" (http://www.stat.gov.pl/broker/access/prefile/listPreFiles.jspa) (in Polish). 2008-06-01. .

Bylin

Bylin	
— Village —	
 Bylin	
Coordinates: 52°20′N 17°8′E	
Country	Poland
Voivodeship	Greater Poland
County	Poznań County
Gmina	Kleszczewo

Bylin [ˈbɨlin] is a village in the administrative district of Gmina Kleszczewo, within Poznań County, Greater Poland Voivodeship, in west-central Poland.[1] It lies approximately 17 km (11 mi) south-east of the regional capital Poznań.

References

[1] "Central Statistical Office (GUS) - TERYT (National Register of Territorial Land Apportionment Journal)" (http://www.stat.gov.pl/broker/access/prefile/listPreFiles.jspa) (in Polish). 2008-06-01. .

Village

A **village** is a clustered human settlement or community, larger than a hamlet with the population ranging from a few hundred to a few thousand (sometimes tens of thousands), Though often located in rural areas, the term urban village is also applied to certain urban neighbourhoods, such as the West Village in Manhattan, New York City and the Saifi Village in Beirut, Lebanon, as well as Hampstead Village in the London conurbation. Villages are normally permanent, with fixed dwellings; however, transient villages[1] can occur. Further, the dwellings of a village are fairly close to one another, not scattered broadly over the landscape, as a dispersed settlement.

Dartlo - the fortified village of Tusheti, Caucasus, Georgia.

Historically, villages were a usual form of community for societies that practise subsistence agriculture, and also for some non-agricultural societies. In Great Britain, a hamlet earned the right to be called a village when it built a church.[2] In many cultures, towns and cities were few, with only a small proportion of the population living in them. The Industrial Revolution attracted people in larger numbers to work in mills and factories; the concentration of people caused many villages to grow into towns and cities. This also enabled specialization of labor and crafts, and development of many trades. The trend of urbanization continues, though not always in connection with industrialisation. Villages have been eclipsed in importance as units of human society and settlement.

Berber village in Ourika valley, High Atlas, Morocco.

Traditional villages

Although many patterns of village life have existed, the typical village was small, consisting of perhaps 5 to 30 families. Homes were situated together for sociability and defence, and land surrounding the living quarters was farmed. Traditional fishing villages were based on artisan fishing and located adjacent to fishing grounds.

The village of Rougon in the Department of Alpes-de-Haute-Provence, France

South Asia

India "The soul of India lives in its villages", declared M. K. Gandhi[3] at the beginning of 20th century. According to the 2001 Indian census, 74% of Indians live in 638,365 different villages.[4] The size of these villages varies considerably. 236,004 Indian villages have a population less than 500, while 3,976 villages have a population of 10,000+. Most of the villages have their own temple, mosque or church depending on the local religious following.

Bangladeshi Village

Masouleh village, Gilan Province, Iran

An alpine village in the Lötschental Valley, Switzerland

Cumalıkızık, a traditional Turkish village, Bursa, Turkey

A village in central India.

East Asia

Shirakawa-gō, Nagano Japan.

People's Republic of China In mainland China, villages are divisions under township Zh:or town Zh:.

Republic of China (Taiwan) In the Republic of China (Taiwan), villages are divisions under townships or county-controlled cities. The village is called a *tsuen* or *cūn* () under a rural township () and a *li* () under an urban township () or a county-controlled city.

Japan

South Korea

Southeast Asia

Brunei, Indonesia, Malaysia and Singapore

The *nagari* of Pariangan, West Sumatra.

In Indonesia, depending on the principles they are administered, villages are called *desa* or *kelurahan.* A *desa* (a term that derives from a Sanskrit word meaning "country" that is found in a name such as "Bangladesh") is administered according to traditions and customary law (*adat*), while a *kelurahan* is administered along more "modern" principles. *Desa* are generally located in rural areas while *kelurahan* are generally urban subdivisions. A village head is respectively called *kepala desa* or *lurah.* Both are elected by the local community. A *desa* or *kelurahan* is itself the subdivision of a *kecamatan* (district), in turn the subdivision of a *kabupaten* (regency).

The same general concept applies all over Indonesia. How ever, there is some variation among the vast numbers of Austronesian ethnic groups. For instance, in Bali villages have been created by grouping traditional hamlets or *banjar*, which constitute the basis of Balinese social life. In the Minangkabau country in West Sumatra province traditional villages are called *nagari* (a term deriving from another Sanskrit word meaning "city", which can be found in a name like

"Srinagar"). In some areas such as Tanah Toraja, elders take turns watching over the village at a command post. As a general rule, *desa* and *kelurahan* are groupings of hamlets (*kampung* in Indonesian, *dusun* in the Javanese language, *banjar* in Bali).

A kampung-type house in Sungai Nipah, Perak, Malaysia

In Malaysia, the term *kampung* (sometimes spelling *kampong*) in the English language has been defined specifically as "a Malay hamlet or village in a Malay-speaking country".[5] In other words, a ***kampung*** is defined today as a village in Brunei, Indonesia or Malaysia. In Malaysia, a *kampung* is determined as a locality with 10,000 or fewer people. Since historical times, every Malay village came under the leadership of a *penghulu* (village chief), who has the power to hear civil matters in his village (see Courts of Malaysia for more details). A Malay village typically contains a *"masjid"* (mosque) or *"surau"* (Muslim chapel), paddy fields and Malay houses on stilts. Malay and Indonesian villagers practice the culture of helping one another as a community, which is better known as "joint bearing of burdens" (*gotong royong*),[6] as well as being family-oriented (especially the concept of respecting one's family [particularly the parents and elders]), courtesy and believing in God (*"Tuhan"*) as paramount to everything else. It is common to see a cemetery near the mosque, as all Muslims in the Malay or Indonesian village want to be prayed for, and to receive Allah's blessings in the afterlife. In Sarawak, some villages are called 'long' pronounced as 'long' in Chinese. These villages are mostly found in western Sarawak.

Singapore also follows the Malaysian *kampung*. However, there are only a few *kampung* villages remaining, mostly on islands surrounding Singapore such as Pulau Ubin. In the past, there was many *kampung* villages in Singapore but now there aren't many on the mainland.

Philippines

In urban areas of the Philippines, the term "village" most commonly refers to private subdivisions, especially gated communities. These villages emerged in the mid-20th century and were initially the domain of elite urban dwellers. Those are common in major cities in the country and their residents have a wide range of income levels. Such villages may or may not correspond to administrative units (usually barangays) and/or be privately administered. Barangays more correspond to the villages of old times, and the chairman (formerly a village datu) now settles intrapersonal matters or polices the village, though with much less authority and respect than in Indonesia or Malaysia.

Vietnam

Village, or "làng", is a basis of Vietnam society. Vietnam's village is the typical symbol of Asian agricultural production. Vietnam's village typically contains: a village gate, "lũy tre" (bamboo hedges), "đình làng" (communal house) where "thành hoàng" (tutelary god) is worshiped, a common well, "đồng lúa" (rice field), "chùa" (temple) and houses of all families in the village. All the people in Vietnam's villages usually have a blood relationship. They are farmers who grow rice and have the same traditional handicraft. Vietnam's villages have an important role in society (Vietnamese saying: "Custom rules the law" -"Phép vua thua lệ làng" [literally: the king's law yields to village customs]). Everyone in Vietnam wants to be buried in their village when they die.

Central and Eastern Europe

Slavic countries

Selo (Cyrillic: село; Polish: *wieś, sioło*) is a Slavic word meaning "village" in Bosnia and Herzegovina, Bulgaria, Croatia, Macedonia, Russia, Serbia, and Ukraine. For example there are numerous *sela* (plural of *selo*) called Novo Selo in Bulgaria, Croatia, Montenegro and others in Serbia, and Macedonia. In Slovenia, the word *selo* is used for very small villages (less than a thousand people) and in dialects; the Slovene word *vas* is used all over Slovenia.

Bulgaria

Kovachevitsa, a village in southern Bulgaria

In Bulgaria, the different types of *Sela* vary from a small selo of 5 to 30 families to one of several thousand people. According to a 2002 census, in that year there were 2,385,000 Bulgarian citizens living in settlements classified as *villages*.[7] A 2004 Human Settlement Profile on Bulgaria[8] conducted by the United Nations Department of Economic and Social Affairs stated that:

> The most intensive is the migration "city – city". Approximately 46% of all migrated people have changed their residence from one city to another. The share of the migration processes "village – city" is significantly less – 23% and "city – village" – 20%. The migration "village – village" in 2002 is 11%.[7]

It also stated that

> the state of the environment in the small towns and villages is good apart from the low level of infrastructure.[7]

In Bulgaria, it is becoming popular to visit villages for the atmosphere, culture, crafts, hospitality of the people and the surrounding nature. This is called *selski tourism* (Bulgarian: селски туризъм), meaning "village tourism".

Russia

The village of Zaponorye in Moscow Oblast

In Russia, as of the 2010 Census, 26.3% of the country's population lives in rural localities;[9] down from 26.7% recorded in the 2002 Census.[9] Multiple types of rural localities exist, but the two most common are *derevnya* (деревня) and *selo* (село). Historically, the formal indication of status was religious: a city (*gorod*) had a cathedral, a *selo* had a church, while a *derevnya* had neither.

The lowest administrative unit of the Russian Empire, a *volost*, or its Soviet or modern Russian successor, a *selsoviet*, was typically headquartered in a *selo* and embraced a few neighboring villages.

Between 1926 and 1989, Russia's rural population shrank from 76 million people to 39 million, due to urbanization, collectivization, dekulakization, and the World War II losses, but has nearly stabilized since. During 1930–1937, mass starvation in Russia and other parts of the Soviet Union lead to the death of at least 14.5 million peasants (including 5-7 million in the Holodomor).[10]

Most Russian rural localities have populations of less than 200 people, and the smaller places take the brunt of depopulation: e.g., in 1959, about one half of Russia's rural population lived in villages of fewer than 500 people, while now less than one third does. In the 1960s–1970s, the depopulation of the smaller villages was driven by the central planners' drive to get the farm workers out of smaller, "prospect-less" hamlets and into the collective or state farms' main villages, with more amenities.[11]

Most Russian rural residents are involved in agricultural work, and it is very common for villagers to produce their own food. As prosperous urbanites purchase village houses for their second homes, Russian villages sometimes are transformed into dacha settlements, used mostly for seasonal residence.

The historically Cossack regions of Southern Russia and parts of Ukraine, with their fertile soil and absence of serfdom, had a rather different pattern of settlement from central and northern Russia. While peasants of central Russia lived in a village around the lord's manor, a Cossack family often lived on its own farm, called *khutor*. A number of such *khutors* plus a central village made up the administrative unit with a center in a *stanitsa* (Russian: стани́ца; Ukrainian: станиця, *stanytsia*). Such *stanitsas*often with a few thousand residents, were usually larger than a typical *selo* in central Russia.

The term *aul/aal* is used to refer mostly Muslim-populated villages in Caucasus and Idel-Ural, without regard to the number of residents.

Ukraine

In Ukraine, a village, known locally as a *"selo"* (село), is considered the lowest administrative unit. Villages may have an individual administration (*silrada*) or a joint administration, combining two or more villages. Villages may also be under the jurisdiction of a city council (*miskrada*) or town council (*selyshchna rada*) administration.

There is, however, another smaller type of settlement which is designated in Ukrainian as a *selysche* (селище). This type of community is generally referred to in English as a "settlement". In comparison with an urban-type settlement, Ukrainian legislation does not have a concrete definition or a criterion to differentiate such settlements from villages. They represent a type of a small rural locality that might have once been a *khutir*, a fisherman's settlement, or a dacha. They are administered by a *silrada* (council) located in a nearby adjacent village. Sometimes the term *"selysche"* is also used in a more general way to refer to adjacent settlements near a bigger city, including urban-type settlements (*selysche miskoho typu*) and/or villages; however, ambiguity is often avoided in connection with urbanized settlements by referring to them using the three-letter abbreviation *smt* instead.

The *khutir* (хутір) and *stanytsia* (станиця) are not part of the administrative division any longer, primarily due to collectivization. *Khutirs* were very small rural localities consisting of just few housing units and were sort of individual farms. They became really popular during the Stolypin reform in the early 20th century. During the collectivization, however, residents of such settlements were usually declared to be kulaks and had all their property confiscated and distributed to others (nationalized) without any compensation. The *stanitsa* likewise has not survived as an administrative term. The *stanitsa* was a type of a collective community that could include one or more settlements such as villages, *khutirs*, and others. Today, *stanitsa*-type formations have only survived in Kuban (Russian Federation) where Ukrainians were resettled during the time of the Russian Empire.

Western & Southern Europe

United Kingdom

A village in the UK is a compact settlement of houses, smaller in size than a town, and generally based on agriculture or, in some areas, mining, such as Ouston, County Durham, quarrying or sea fishing.

The major factors in the type of settlement are location of water sources, organisation of agriculture and landholding, and likelihood of flooding. For example, in areas such as the Lincolnshire Wolds, the villages are often found along the spring line halfway down the hillsides, and originate as spring line settlements, with the original open field systems around the village. In northern Scotland, most villages are planned to a grid pattern located on or close to major roads, whereas in areas such as the Forest of Arden, woodland clearances produced small hamlets around village greens.[12] [13]

The main street of the village of Castle Combe, Wiltshire, England.

Some villages have disappeared (for example, deserted medieval villages), sometimes leaving behind a church or manor house and sometimes nothing but bumps in the fields.Some show archaeological evidence of settlement at three or four different layers, each distinct from the previous one. Clearances may have been to accommodate sheep or game estates, or enclosure, or may have resulted from depopulation, such as after the Black Death or following a move of the inhabitants to more prosperous districts. Other villages have grown and merged and often form hubs within the general mass of suburbia — such as Hampstead, London and Didsbury in Manchester. Many villages are now predominantly dormitory locations and have suffered the loss of shops, churches and other facilities.

For many British people, the village represents an ideal of Great Britain. Seen as being far from the bustle of modern life, it is represented as quiet and harmonious, if a little inward-looking. This concept of an unspoilt Arcadia is present in many popular representations of the village such as the radio serial *The Archers* or the best kept village competitions. The reality is that many villages are plagued by lack of access to public transport and local services, specially affecting the poor and elderly who cannot afford their own means of transport.[14]

Many villages in South Yorkshire, North Nottinghamshire, North East Derbyshire, County Durham, South Wales and Northumberland are known as pit villages. These (such as Murton, County Durham) grew from hamlets when the sinking of a colliery in the early 20th century resulted in a rapid growth in their population and the colliery owners built new housing, shops, pubs and churches. Some pit villages outgrew nearby towns by area and population; for example, Rossington in South Yorkshire came to have over four times more people than the nearby town of Bawtry. Some pit villages grew to become towns; for example, Maltby in South Yorkshire grew from 600 people in the 19th century[15] to over 17,000 in 2007.[16] Maltby was constructed under the auspices of the Sheepbridge Coal and Iron Company and included ample open spaces and provision for gardens.[17]

Bisley, Gloucestershire, a village in the Cotswolds

In the UK, the main historical distinction between a hamlet and a village was that the latter had a church,[2] and so usually was the centre of worship for an ecclesiastical parish. However, some civil parishes may contain more than one village. The typical village had a pub or inn, shops, and a blacksmith. But many of these facilities are now gone, and many villages are dormitories for commuters. The population of such settlements ranges from a few hundred people to around five thousand. A village is distinguished from a town in that:

- A village should not have a regular agricultural market, although today such markets are uncommon even in settlements which clearly are towns.
- A village does not have a town hall nor a mayor.
- If a village is the principal settlement of a civil parish, then any administrative body that administers it at parish level should be called a parish council or parish meeting, and not a town council or city council. However, some civil parishes have no functioning parish, town, or city council nor a functioning parish meeting. In Wales, where the equivalent of an English civil parish is called a Community, the body that administers it is called a Community Council. However, larger councils may elect to call themselves town councils.[18] Unlike Wales, Scottish community councils have no statutory powers.[19]
- There should be a clear green belt or open fields, as, for example, seen on aerial maps for Ouston surrounding its parish[20] borders. However this may not be applicable to urbanised villages: although these may not considered to be villages, they are often widely referred to as being so; an example of this is Horsforth in Leeds.

France

Same general definition as in the UK.

An independent association named *Les Plus Beaux Villages de France*, was created in 1982 to promote assets of small and picturesque French villages of quality heritage. As of 2008, 152 villages in France have been labelled as "The Most Beautiful Villages of France".

Saint-Cirq-Lapopie (Lot) is one of "The Most Beautiful Villages in France".

Spain

Spain has plenty of little villages around its territory. Country life is more usual in Castile and Aragon. All villages have a church or hermitage.

Portugal

Villages are more usual in the northern and central regions and in the Alentejo. Most of them have a church and a "Casa do Povo" (people's house), where the village's summer **romarias** or religious festivities are usually held. Summer is also when many villages are host to a range of folk festivals and fairs, taking advantage of the fact that many of the locals who reside abroad tend to come back to their native village for the holidays.

Netherlands

In the flood prone districts of the Netherlands, villages were traditionally built on low man-made hills called terps before the introduction of regional dyke-systems. In modern days, the term *dorp* (lit. "village") is usually applied to settlements no larger than 20,000, though there's no official law regarding status of settlements in the Netherlands.

Middle East

Lebanon

The main square of Saifi Village in Centre Ville, Beirut, Lebanon

Like France, villages in Lebanon are usually located in remote mountainous areas. The majority of villages in Lebanon retain their Aramaic names or are derivative of the Aramaic names, and this is because Aramaic was still in use in Mount Lebanon up to the 18th century.[21]

Many of the Lebanese villages are a part of districts, these districts are known as "kadaa" which includes the districts of Baabda (Baabda), Aley (Aley), Matn (Jdeideh), Keserwan (Jounieh), Chouf (Beiteddine), Jbeil (Byblos), Tripoli (Tripoli), Zgharta (Zgharta / Ehden), Bsharri (Bsharri), Batroun (Batroun), Koura (Amioun), Miniyeh-Danniyeh (Minyeh / Sir Ed-Danniyeh), Zahle (Zahle), Rashaya (Rashaya), Western Beqaa (Jebjennine / Saghbine), Sidon (Sidon), Jezzine (Jezzine), Tyre (Tyre), Nabatiyeh (Nabatiyeh), Marjeyoun (Marjeyoun), Hasbaya (Hasbaya), Bint Jbeil (Bint Jbeil), Baalbek (Baalbek), and Hermel (Hermel).

The district of Danniyeh consists of thirty six small villages, which includes Almrah, Kfirchlan, Kfirhbab, Hakel al Azimah, Siir, Bakhoun, Miryata, Assoun, Sfiiri, Kharnoub, Katteen, Kfirhabou, Zghartegrein, Ein Qibil.

Danniyeh (known also as Addinniyeh, Al Dinniyeh, Al Danniyeh, Arabic: سير الضنية) is a region located in Miniyeh-Danniyeh District in the North Governorate of Lebanon. The region lies east of Tripoli, extends north as far as Akkar District, south to Bsharri District and Zgharta District and as far east as Baalbek and Hermel. Dinniyeh has an excellent ecological environment filled with woodlands, orchards and groves. Several villages are located in this mountainous area, the largest town being Sir Al Dinniyeh.

An example of a typical mountainous Lebanese village in Dannieh would be Hakel al Azimah which is a small village that belongs to the district of Danniyeh, situated between Bakhoun and Assoun's boundaries. It is in the centre of the valleys that lie between the Arbeen Mountains and the Khanzouh.

Syria

General view from Al-Annaze village, near Tartus, Syria

Syria contains a large number of villages that vary in size and importance, including the ancient, historical and religious villages, such as Ma'loula, Sednaya, and Brad (Mar Maroun's time). The diversity of the Syrian environments creates significant differences between the Syrian villages in terms of the economic activity and the method of adoption. Villages in the south of Syria (Huran, Jabal Al-Arab), the north-east (the Syrian island) and the Orontes River basin depend mostly on agriculture, mainly grain, vegetables and fruits. Villages in the region of Damascus and Aleppo depend on trading. Some other villages, such as Marmarita depend heavily on tourist activity.

Mediterranean cities in Syria, such as Tartus and Latakia have similar types of villages. Mainly, villages were built in very good sites which had the fundamentals of the rural life, like water. An example of a Mediterranean Syrian village in Tartus would be Al-Annaze, which is a small village that belongs to the area of Al Sauda. The area of Al Sauda is called a nahiya, which is a subdistrict.

Australasia & Oceania

The village of Puamau on Hiva Oa, Marquesas Islands, French Polynesia

Pacific Islands Communities on pacific islands were historically called villages by English speakers who traveled and settled in the area. Some communities such as several Villages of Guam continue to be called villages despite having large populations that can exceed 40,000 residents.

New Zealand The traditional Māori village was the pā, a fortified hill-top settlement. Tree-fern logs and flax were the main building materials.

Australia The term village often is used in reference to small planned communities such as retirement communities or shopping districts, and tourist areas such as ski resorts. Small rural communities are usually known as townships. Larger settlements are known as towns.

South America

Argentina Usually set in remote mountainous areas, some also cater to winter sports and/or tourism, see: Uspallata, La Cumbrecita, Villa Traful and La Cumbre

North America

Canada

United States

A Newfoundland fishing village

Incorporated villages

In twenty[22] U.S. states, the term "village" refers to a specific form of incorporated municipal government, similar to a city but with less authority and geographic scope. However, this is a generality; in many states, there are villages that are an order of magnitude larger than the smallest cities in the state. The distinction is not necessarily based on population, but on the relative powers granted to the different types of municipalities and correspondingly, different obligations to provide specific services to residents.

In some states such as New York, Wisconsin, or Michigan, a village is an incorporated municipality, usually, but not always, within a single town or civil township. Residents pay taxes to the village and town or township and may vote in elections for both as well. In some cases, the village may be coterminous with the town or township. There are also many villages which span the boundaries of more than one town or township, and some villages may even straddle county borders.

There is no limit to the population of a village in New York; Hempstead, the largest village in the state, has 55,000 residents, making it more populous than some of the state's cities. However, villages in the state may not exceed five square miles (13 km²) in area.

In the state of Wisconsin, a village is always legally separate from the towns that it has been incorporated from. The largest village is Menomonee Falls, which has over 32,000 residents.

Michigan and Illinois also have no set population limit for villages and there are many villages that are larger than cities in those states. The village of Arlington Heights, IL had 75,101 residents as of the 2010 census.

Villages in Ohio are often legally part of the township from which they were incorporated, although exceptions such as Hiram exist, in which the village is separate from the township.[23] They have no area limitations, but become cities if they grow a population of more than 5,000.[24]

In Maryland, a locality designated "Village of ..." may be either an incorporated town or a special tax district.[25] An example of the latter is the Village of Friendship Heights.

In states that have New England towns, a "village" is a center of population or trade, including the town center, in an otherwise sparsely-developed town or city — for instance, the village of Hyannis in the city of the Barnstable, Massachusetts.

Unincorporated villages

In many states, the term "village" is used to refer to a relatively small unincorporated community, similar to a hamlet in New York state. This informal usage may be found even in states that have villages as an incorporated municipality, although such usage might be considered incorrect and confusing.

See also

- Global village
- Linear village
- Village green
- Village lock-up
- police village

Settlement types

- Dugout
- Fishing village
- Hamlet
- Microtown

Countries and localities

- Dhani and villages
- Dogon villages
- Hakka architecture
- Ksar
- List of villages in Europe by country
- Pueblo
- Sołectwo (rough equivalent in Poland)
- Ville

Developed environments

- Developed environments
- City
- Exurban
- Megalopolis
- Rural
- Suburban
- Urban area

Footnotes

[1] http://www.google.co.uk/search?hl=en&safe=off&q=%22transient+villages%22&btnG=Search&meta=

[2] Dr Greg Stevenson, "What is a Village?" (http://www.bbc.co.uk/history/programmes/restoration/2006/exploring_brit_villages_01.shtml), *Exploring British Villages*, BBC, 2006, accessed 20 October 2009

[3] http://www.pibbng.kar.nic.in/feature1.pdf

[4] "Indian Census" (http://www.censusindia.gov.in/). Censusindia.gov.in. . Retrieved 2010-03-28.

[5] "Meriam-Webster Online" (http://www.m-w.com/dictionary/kampung). M-w.com. 2007-04-25. . Retrieved 2010-03-28.

[6] Geertz, Clifford. "Local Knowledge: Fact and Law in Comparative Perspective", pp. 167-234 in Geertz *Local Knowledge: Further Essays in Interpretive Anthropology,* NY: Basic Books. 1983.

[7] "Human Settlement Country Profile, Bulgaria (*2004*)" (http://www.un.org/esa/agenda21/natlinfo/countr/bulgaria/Bulgariahumansettlement2003.PDF) (PDF). United Nations Department of Economic and Social Affairs. . Retrieved 2008-11-30.

[8] http://www.un.org/esa/agenda21/natlinfo/countr/bulgaria/Bulgariahumansettlement2003.PDF

[9] Федеральная служба государственной статистики (Federal State Statistics Service) (2011). "Предварительные итоги Всероссийской переписи населения 2010 года[[Category:Articles containing non-English language text (http://www.perepis-2010.ru/results_of_the_census/results-inform.php)] (*Preliminary results of the 2010 All-Russian Population Census*)"] (in Russian). *Всероссийская перепись населения 2010 года (All-Russia Population Census of 2010).* Federal State Statistics Service. . Retrieved 2011-04-25.

[10] Robert Conquest (1986) *The Harvest of Sorrow: Soviet Collectivization and the Terror-Famine*. Oxford University Press. ISBN 0-19-505180-7.

[11] "Российское село в демографическом измерении" (*Rural Russia measured demographically*) (http://demoscope.ru/weekly/2006/0253/tema04.php) (Russian). This article reports the following census statistics:

Census year	1959	1970	1979	1989	2002
Total number of rural localities in Russia	294,059	216,845	177,047	152,922	155,289
Of them, with population 1 to 10 persons	41,493	25,895	23,855	30,170	47,089
Of them, with population 11 to 200 persons	186,437	132,515	105,112	80,663	68,807

[12] Wild, Martin Trevor (2004). *Village England: a social history of the countryside* (http://books.google.co.uk/books?id=M7AmyuGr5Y8C&printsec=frontcover). I.B.Tauris. p. 12. ISBN 9781860649394. .

[13] Taylor, Christopher (1984). *Village and farmstead: A history of rural settlement in England* (http://books.google.co.uk/books?ei=NksHTpXqIo6t8QPE1ozADQ). G. Philip. p. 192. ISBN 9780540010820. .

[14] OECD (2011). *OECD Rural Policy Reviews: England, United Kingdom 2011* (http://books.google.co.uk/books?id=bQCGWKfXJNMC&printsec=frontcover&source=gbs_ge_summary_r&cad=0). OECD Publishing. p. 237. .

[15] *The Parliamentary gazetteer of England and Wales* (http://books.google.co.uk/books?id=mxIQAAAAYAAJ&printsec=frontcover). **3**. A. Fullarton & Co.. 1851. p. 344. .

[16] "Maltby Ward" (http://www.rotherham.gov.uk/download/553/maltby_ward). Rotherham Metropolitan Borough Council. . Retrieved 2011-06-26.

[17] Baylies, Carolyn Louise (1993). *The history of the Yorkshire miners, 1881-1918* (http://books.google.co.uk/books?id=WEIOAAAAQAAJ&printsec=frontcover&source=gbs_ge_summary_r&cad=0). Routledge. .

[18] "National Statistics" (http://www.statistics.gov.uk/geography/parishes.asp). Statistics.gov.uk. . Retrieved 2010-03-28.

[19] "Portobello Community Council" (http://www.porty.org.uk/council/index.php). Porty.org.uk. . Retrieved 2010-03-28.

[20] "Ouston Parish Council" (http://parishes.durham.gov.uk/ouston/Pages/wherewelive.aspx). durham.gov.uk. .

[21] "A project proposal" (http://almashriq.hiof.no/lebanon/400/410/412/elies_project/glimse_of_yesterday.html). Almashriq.hiof.no. . Retrieved 2010-03-28.

[22] "Village" (http://www.websters-online-dictionary.org/definition/english/vi/village.html#Definitions). Websters-online-dictionary.org. . Retrieved 2010-03-28.

[23] "Detailed map of Ohio" (http://www2.census.gov/geo/maps/general_ref/cousub_outline/cen2k_pgsz/oh_cosub.pdf) (PDF). United States Census Bureau. 2000. . Retrieved 2010-03-28.

[24] "Ohio Revised Code Section 703.01(A)" (http://codes.ohio.gov/orc/703.01). . Retrieved 2010-03-28.

[25] 2002 Census of Governments, Individual State Descriptions (http://www.census.gov/prod/2005pubs/gc021x2.pdf) (PDF)

External links

- Types of villages (anthropogenic biomes) (http://www.ecotope.org/anthromes/v1/guide/villages/)

bjn:Disa gag:Küü rue:Село

Gmina_Kórnik

Gmina Kórnik Kórnik Commune	
— Gmina —	
Coordinates (Kórnik): 52°14′12″N 17°5′55″E	
Country	Poland
Voivodeship	Greater Poland
County	Poznań County
Seat	Kórnik
Area	
• **Total**	186.58 km^2 (72 sq mi)
Population (2006)	
• **Total**	17585
• **Density**	94.2/km^2 (244.1/sq mi)
• **Urban**	6981
• **Rural**	10604
Website	http://www.kornik.pl/

Gmina Kórnik is an urban-rural gmina (administrative district) in Poznań County, Greater Poland Voivodeship, in west-central Poland. Its seat is the town of Kórnik, which lies approximately 22 kilometres (14 mi) south-east of the regional capital Poznań.

The gmina covers an area of 186.58 square kilometres (72.0 sq mi), and as of 2006 its total population is 17,585 (out of which the population of Kórnik amounts to 6,981, and the population of the rural part of the gmina is 10,604).

Villages

Apart from the town of Kórnik, Gmina Kórnik contains the villages and settlements of Biernatki, Błażejewko, Błażejewo, Borówiec, Czmoń, Czmoniec, Czołowo, Dachowa, Dębiec, Dworzyska, Dziećmierowo, Gądki, Jaryszki, Kamionki, Konarskie, Koninko, Mościenica, Pierzchno, Prusinowo, Radzewo, Robakowo, Runowo, Skrzynki, Świątniczki, Szczodrzykowo, Szczytniki, Trzykolne Młyny and Żerniki.

Neighbouring gminas

Gmina Kórnik is bordered by the city of Poznań and by the gminas of Kleszczewo, Mosina, Śrem, Środa Wielkopolska and Zaniemyśl.

References

- Polish official population figures 2006 [1]

References

[1] http://www.stat.gov.pl/gus/45_655_PLK_HTML.htm

Poznań_County

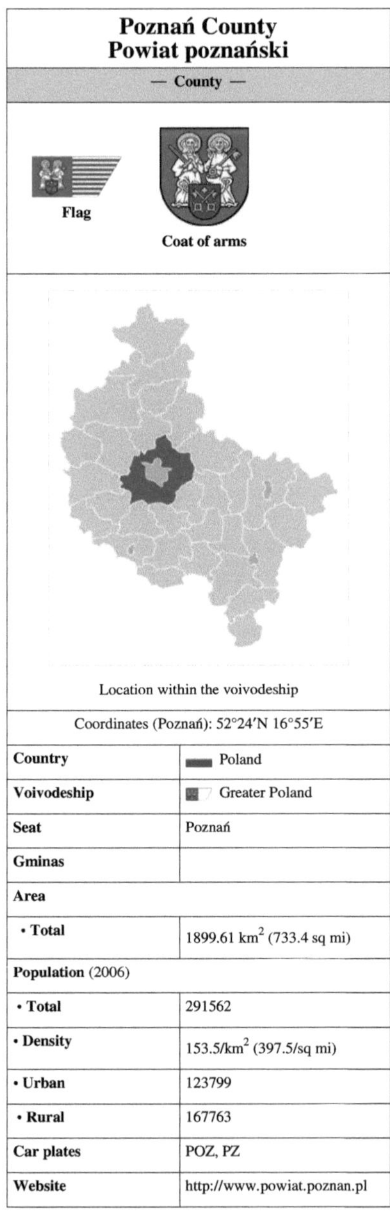

Poznań County Powiat poznański	
— County —	
Flag / Coat of arms	
Location within the voivodeship	
Coordinates (Poznań): 52°24′N 16°55′E	
Country	Poland
Voivodeship	Greater Poland
Seat	Poznań
Gminas	
Area	
• **Total**	1899.61 km^2 (733.4 sq mi)
Population (2006)	
• **Total**	291562
• **Density**	153.5/km^2 (397.5/sq mi)
• **Urban**	123799
• **Rural**	167763
Car plates	POZ, PZ
Website	http://www.powiat.poznan.pl

Poznań County (Polish: *powiat poznański*) is a unit of territorial administration and local government (powiat) in Greater Poland Voivodeship, west-central Poland. It came into being on January 1, 1999, as a result of the Polish local government reforms passed in 1998. Its administrative seat is the city of Poznań, although the city is not part of the county (it constitutes a separate city county). The county's administrative offices (*starostwo powiatowe*) are in the Jeżyce neighbourhood of Poznań.

Poznań County contains 10 towns: Swarzędz, 11 km (7 mi) east of (central) Poznań, Luboń, 8 km (5 mi) south of Poznań, Mosina, 18 km (11 mi) south of Poznań, Murowana Goślina, 20 km (12 mi) north of Poznań, Puszczykowo, 14 km (9 mi) south of Poznań, Kostrzyn, 21 km (13 mi) east of Poznań, Pobiedziska, 27 km (17 mi) north-east of Poznań, Kórnik, 22 km (14 mi) south-east of Poznań, Buk, 28 km (17 mi) west of Poznań, and Stęszew, 21 km (13 mi) south-west of Poznań.

The county covers an area of 1899.61 square kilometres (733.4 sq mi). As of 2006 its total population is 291,562, out of which the urban population is 123,799 (Swarzędz 29,894, Luboń 26,935, Mosina 12,150, Murowana Goślina 10,140, Puszczykowo 9,311, Kostrzyn 8,539, Pobiedziska 8,329, Kórnik 6,981, Buk 6,181, Stęszew 5,339), and the rural population is 167,763.

Neighbouring counties

Apart from the city of Poznań, Poznań County is also bordered by Oborniki County and Wągrowiec County to the north, Gniezno County and Września County to the east, Środa Wielkopolska County to the south-east, Śrem County and Kościan County to the south, Grodzisk Wielkopolski County and Nowy Tomyśl County to the west, and Szamotuły County to the north-west.

Administrative division

The county is subdivided into 17 gminas (two urban, eight urban-rural and seven rural). These are listed in the following table, in descending order of population.

Gmina	Type	Area (km^2)	Population (2006)	Seat
Gmina Swarzędz	urban-rural	102.0	40,499	Swarzędz
Luboń	urban	13.5	26,935	
Gmina Mosina	urban-rural	170.9	25,098	Mosina
Gmina Czerwonak	rural	82.2	23,692	Czerwonak
Gmina Tarnowo Podgórne	rural	101.4	18,690	Tarnowo Podgórne
Gmina Kórnik	urban-rural	186.6	17,585	Kórnik
Gmina Pobiedziska	urban-rural	189.3	16,382	Pobiedziska
Gmina Murowana Goślina	urban-rural	172.1	15,766	Murowana Goślina
Gmina Kostrzyn	urban-rural	154.0	15,456	Kostrzyn
Gmina Komorniki	rural	66.6	14,353	Komorniki
Gmina Stęszew	urban-rural	175.2	13,919	Stęszew
Gmina Dopiewo	rural	108.1	13,889	Dopiewo
Gmina Suchy Las	rural	116.6	13,219	Suchy Las
Gmina Buk	urban-rural	90.3	11,917	Buk
Gmina Rokietnica	rural	79.3	9,415	Rokietnica
Puszczykowo	urban	16.7	9,311	
Gmina Kleszczewo	rural	74.8	5,436	Kleszczewo

Map

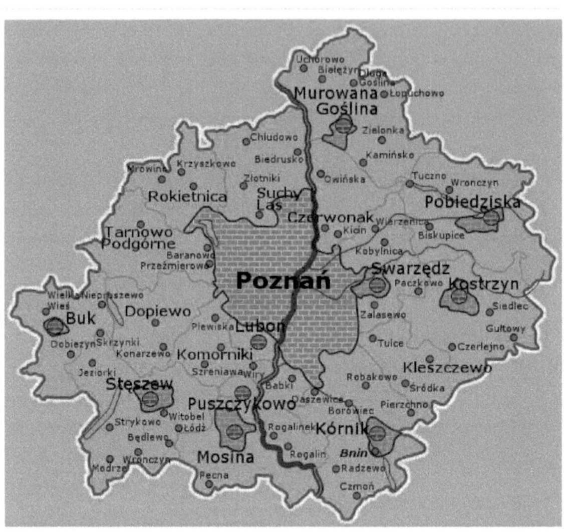

A map of Poznań County. The pink lines indicate the boundaries of towns and gminas.

References

- Polish official population figures 2006 [1]

Greater_Poland_Voivodeship

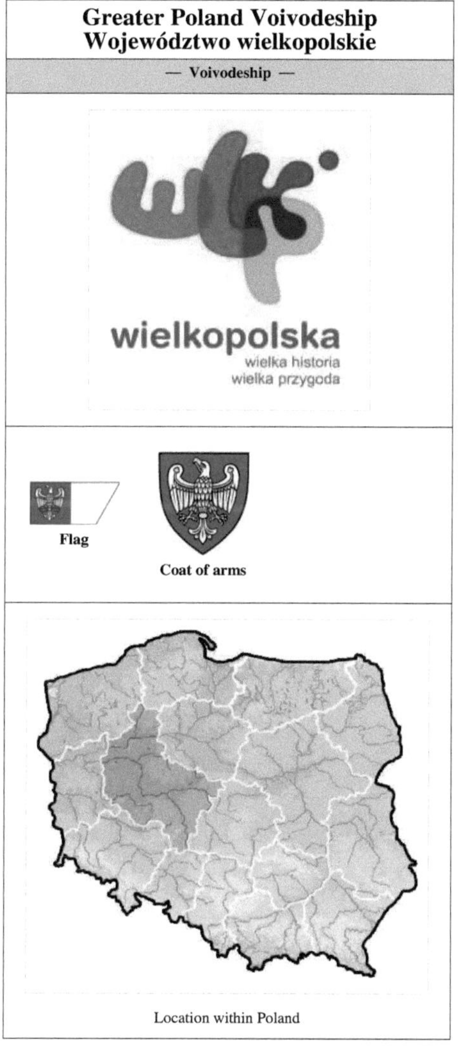

Flag

Coat of arms

Location within Poland

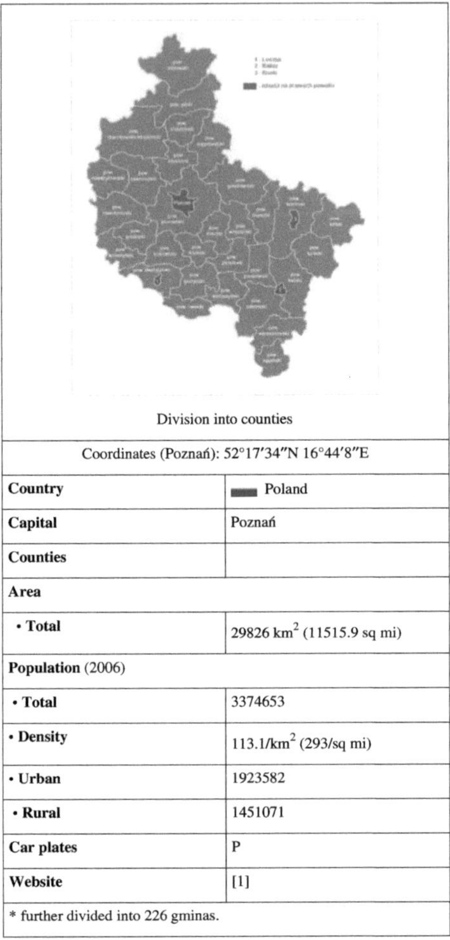

Division into counties

Coordinates (Poznań): 52°17′34″N 16°44′8″E	
Country	Poland
Capital	Poznań
Counties	
Area	
• **Total**	29826 km^2 (11515.9 sq mi)
Population (2006)	
• **Total**	3374653
• **Density**	113.1/km^2 (293/sq mi)
• **Urban**	1923582
• **Rural**	1451071
Car plates	P
Website	[1]
* further divided into 226 gminas.	

Wielkopolska Voivodeship (in Polish, *województwo wielkopolskie* [vɔjɛˈvut͡stfɔ vjɛlkɔˈpɔlskʲɛ]), or **Greater Poland Voivodeship**, is a voivodeship, or province, in west-central Poland. It was created on 1 January 1999 out of the former Poznań, Kalisz, Konin, Piła and Leszno Voivodeships, pursuant to the Polish local government reforms adopted in 1998. The province is named after the region called Greater Poland or *Wielkopolska* [vjɛlkɔˈpɔlska] (listen). The modern province includes most of this historic region, except for some south-western parts.

Greater Poland Voivodeship is second in area and third in population among Poland's sixteen voivodeships, with an area of 29826 square kilometres (11516 sq mi) and a population of close to 3.4 million. Its capital city is Poznań; other important cities include Kalisz, Konin, Piła, Ostrów Wielkopolski and Gniezno (an early capital of Poland). It is bordered by seven other voivodeships: West Pomeranian to the northwest, Pomeranian to the north, Kuyavian-Pomeranian to the north-east, Łódź to the south-east, Opole to the south, Lower Silesian to the southwest and Lubusz to the west.

Greater Poland Voivodeship has the highest percentage of unemployment, making it prominent among all polish provinces, and the city of Poznań has international twinning arrangements with the English county of Nottinghamshire.[2]

History

Greater Poland, sometimes called the "cradle of Poland," formed the heart of the 10th-century early Polish state. Poznań and Gniezno were early centers of royal power, but following the region's devastation by pagan rebellion in the 1030s, and an invasion by Bretislaus I of Bohemia in 1038, the capital was moved by Casimir the Restorer from Gniezno to Kraków.

Kórnik castle

In the testament of Bolesław III Krzywousty, which initiated the period of fragmentation of Poland (1138–1320), the western part of Greater Poland (including Poznań) was granted to Mieszko III the Old. The eastern part, with Gniezno and Kalisz, was part of the Duchy of Kraków, granted to Władysław II. However for most of the period the two parts were under a single ruler, and were known as the Duchy of Greater Poland (although at times there were separately ruled duchies of Poznań, Gniezno, Kalisz and Ujście). The region came under the control of Władysław I the Elbow-High in 1314, and thus became part of the reunited Poland of which Władyslaw was crowned king in 1320.

In the reunited kingdom, and later in the Polish–Lithuanian Commonwealth, the country came to be divided into administrative units called voivodeships. In the case of the Greater Poland region these were Poznań Voivodeship and Kalisz Voivodeship. The Commonwealth also had larger subdivisions known as *prowincja*, one of which was named Greater Poland. However, this *prowincja* covered a larger area than the Greater Poland region itself, also taking in Masovia and Royal Prussia. (This division of Crown Poland into two entities called Greater and Lesser Poland had its roots in the Statutes of Casimir the Great of 1346–1362, where the laws of "Greater Poland" – the northern part of the country – were codified in the Piotrków statute, with those of "Lesser Poland" in the separate Wiślica statute.)

In 1768 a new Gniezno Voivodeship was formed out of the northern part of Kalisz Voivodeship. However more far-reaching changes would come with the Partitions of Poland. In the first partition (1772), northern parts of Greater Poland along the Noteć (German *Netze*) were taken over by Prussia, becoming the Netze District. In the second partition (1793) the whole of Greater Poland was absorbed by Prussia, becoming part of the province of South Prussia. It remained so in spite of the first Greater Poland Uprising (1794), part of the unsuccessful Kościuszko Uprising directed chiefly against Russia.

More successful was the Greater Poland Uprising of 1806, which led to the region's becoming part of the Napoleonic Duchy of Warsaw (forming the Poznań Department and parts of the Kalisz and Bydgoszcz Departments). However, following the Congress of Vienna in 1815, Greater Poland was again partitioned, with the western part (including Poznań) going to Prussia. The eastern part joined the Russian-controlled Kingdom of Poland, where it formed the Kalisz Voivodeship until 1837, then the Kalisz Governorate (merged into the Warsaw Governorate between 1844 and 1867).

Soldiers in the Greater Poland Uprising of 1918–1919

Within the Prussian empire, western Greater Poland became the Grand Duchy of Posen (Poznań), which theoretically held some autonomy. Following an unrealized uprising in 1846, and the more substantial but still unsuccessul uprising of 1848 (during the Spring of Nations), the Grand Duchy was replaced by the Province of Posen. The authorities made efforts to Germanize the region, particularly after the founding of Germany in 1871, and from 1886 onwards the Prussian Settlement Commission was active in increasing German land ownership in formerly Polish areas.

Following the end of World War I, the Greater Poland Uprising (1918–1919) ensured that most of the region became part of the newly independent Polish state, forming most of Poznań Voivodeship (1921–1939). Northern and some western parts of Greater Poland remained in Germany, where they formed much of the province of Posen-West Prussia (1922–1938), whose capital was Schneidemühl (Piła).

Following the German invasion of 1939, Greater Poland was incorporated into Nazi Germany, becoming the province called Reichsgau Posen, later Reichsgau Wartheland (*Warthe* being the German name for the Warta river). The Polish population was oppressed, with many former officials and others considered potential enemies by the Nazis being imprisoned or executed, including at the notorious Fort VII concentration camp in Poznań. Poznań was declared a stronghold city *(Festung)* in the closing stages of the war, being taken by the Red Army in the Battle of Poznań, which ended on 22 February 1945.

After the war, Greater Poland was fully within the Polish People's Republic, as Poznań Voivodeship. With the reforms of 1975 this was divided into smaller provinces (the voivodeships of Kalisz, Konin, Leszno and Piła, and a smaller Poznań Voivodeship). The present-day Greater Poland Voivodeship, again with Poznań as its capital, was created in 1999.

Cities and towns

The voivodeship contains 109 cities and towns. These are listed below in descending order of population (according to official figures for 2006[3]):

Poznań is the capital of the Greater Poland Voivodeship

1. Poznań (566,546)
2. Kalisz (108,575)
3. Konin (80,618)
4. Piła (75,044)
5. Ostrów Wielkopolski (72,577)
6. Gniezno (70,080)
7. Leszno (64,079)
8. Śrem (30,227)
9. Swarzędz (29,894)
10. Krotoszyn (29,421)
11. Turek (29,302)
12. Września (28,617)
13. Luboń (26,935)
14. Jarocin (25,834)
15. Wągrowiec (24,681)
16. Kościan (24,102)
17. Koło (23,034)

18. Środa Wielkopolska (21,635)
19. Rawicz (21,301)
20. Gostyń (20,588)
21. Chodzież (19,652)
22. Szamotuły (18,760)
23. Złotów (18,468)
24. Oborniki (17,850)
25. Pleszew (17,787)
26. Trzcianka (16,756)
27. Nowy Tomyśl (15,225)
28. Kępno (14,710)
29. Ostrzeszów (14,536)
30. Słupca (14,363)
31. Grodzisk Wielkopolski (13,703)
32. Wolsztyn (13,557)
33. Mosina (12,150)
34. Wronki (11,551)
35. Czarnków (11,356)
36. Międzychód (10,920)
37. Rogoźno (10,905)

Kalisz Old Town Hall

Gniezno with its cathedral is the seat of the catholic Primate of Poland

1. Murowana Goślina (10,140)
2. Puszczykowo (9,311)
3. Opalenica (9,104)
4. Kostrzyn (8,539)
5. Jastrowie (8,403)
6. Pobiedziska (8,329)
7. Witkowo (7,855)
8. Trzemeszno (7,789)
9. Pniewy (7,464)
10. Zbąszyń (7,300)
11. Kórnik (6,981)
12. Kłodawa (6,829)
13. Koźmin Wielkopolski (6,707)
14. Krzyż Wielkopolski (6,283)
15. Buk (6,181)
16. Sieraków (5,994)
17. Wieleń (5,940)
18. Śmigiel (5,452)
19. Stęszew (5,339)
20. Wyrzysk (5,234)
21. Czempiń (5,135)
22. Nowe Skalmierzyce (5,080)
23. Odolanów (4,960)

24. Zduny (4,498)
25. Golina (4,330)
26. Szamocin (4,267)
27. Kleczew (4,173)
28. Krobia (4,022)
29. Ujście (3,899)
30. Skoki (3,866)
31. Okonek (3,827)
32. Sompolno (3,695)
33. Krajenka (3,651)
34. Miłosław (3,589)
35. Tuliszków (3,393)
36. Gołańcz (3,342)
37. Rakoniewice (3,253)

1. Nekla (3,203)
2. Pyzdry (3,188)
3. Łobżenica (3,172)
4. Miejska Górka (3,128)
5. Ślesin (3,102)
6. Kobylin (3,084)
7. Bojanowo (3,014)
8. Margonin (2,956)
9. Zagórów (2,932)
10. Lwówek (2,909)
11. Poniec (2,875)
12. Sulmierzyce (2,772)
13. Wysoka (2,750)
14. Książ Wielkopolski (2,724)
15. Kłecko (2,677)
16. Czerniejewo (2,556)
17. Rydzyna (2,539)
18. Borek Wielkopolski (2,486)
19. Rychwał (2,377)
20. Obrzycko (2,170)
21. Dąbie (2,087)
22. Żerków (2,058)
23. Raszków (2,037)
24. Osieczna (2,018)
25. Ostroróg (1,995)
26. Pogorzela (1,974)
27. Grabów nad Prosną (1,967)
28. Jutrosin (1,872)
29. Mikstat (1,840)
30. Przedecz (1,771)
31. Wielichowo (1,765)
32. Stawiszyn (1,554)
33. Krzywiń (1,547)

34. Dobra (1,511)
35. Dolsk (1,479)

Geography

Topography

Lake Góreckie in Wielkopolska National Park

The relief of Greater Poland, geological conditions and soil have been shaped by two glaciations:

- The Baltic glaciation in the lowlands of northern and central Europe where there are now numerous lakes of the Pomeranian Lake District, a feature especially common in and around Poznań and Gniezno.
- The Mid-glaciation in the southern part of the province, where there is less terrain diversity and a lack of major lakes.

The highest elevation is Greater Kobyla Mountain (284 m) in the Ostrzeszowski Hills, the lowest area is located in the valley of the Warta River at the mouth of its tributary the Noteć (21 m) in the north-western part of the region. Agriculturally fertile soils account for around 60% of the province's area, whilst 20%, the rest of the none forested or urban areas, is mostly wetland soil (muck-peat and alluvial soils).

An area of approximately 800 thousand hectares is covered by forests, this represents around 25.8% of the total surface area of the region. In the lake districts of the northern and central parts of the province there are about 800 lakes; 58% of which cover an area of at least 10 hectares and 8%, with an area exceeding 100 hectares. The largest reservoir is the natural Greater Powidzkie Lake (1036 ha) in the Gniezno Lake District. Wielkopolska Region lies within the basin of the Oder River, 88% of the province's surface water drains into the Warta river basin, whilst the remaining 12% is drained by a multitude of other river systems, including the Barycz, Ladislaus Trench and Obrzycy waterways. The quality of river waters is generally poor, but their condition is gradually improving and should soon be classed as 'clean'.

Geology

A modern coal-fired power plant in Pątnów

The main mineral energy resources in Greater Poland are lignite, natural gas, oil and peat.

Brown coal deposits are currently mined in the Konin area, form the basis for the province's power industry (The Pątnów-Adams-Konin coal-fired power stations account for more than 10% of national electricity production). There are also significant quantities of peat deposits in the region; it is calculated that there are ca. 886 thousand hectares of land covered with an average thickness of 1.5 m of peat. An abundance of raw materials used in the production of numerous medicine was recently discovered in the muds of Błażejewo, Oderbank and Mechnacz. In addition, very large deposits of brown coal have been discovered in the vicinity of Kościan, these however are not currently being extracted and probably due to the expense that would be incurred in adapting the site to build a mine and the need to resettle thousands of people never will be. Rock salt is mined intensively at a salt mine in Kłodawa (this mine alone accounts for about 20% of domestic production). The Lime exploiting rich deposits of gypsum.

Throughout the province there are significant deposits of aggregates, ceramic materials and lacustrine chalk. In Kościan the largest and most modern, Polish natural gas production site is in operation. It supplies raw material for Kościańska Zieme, and Zielona Gora CHP. It is estimated that gas reserves being exploited at the plant in Kościan will be enough for about 20 years of operation, thus practically allowing for local independence against the effects of gas crises.

Climate

Wielkopolska is influenced by oceanic air masses that affect the mildness of the climate. The farther east one travels the more distinctly continental the climate becomes. The area is situated in the Silesian Greater Poland agro-climatic region where the average annual temperature is about 8.2 °C, and in the north drops to around 7.6 °C. It is slightly warmer in the south and west where the average temperature is usually about 8.5 °C. The number of days with snow can reach up to 57 days in and around the Kalisz district.

The growing season is one of the longest in Poland. On the province's southern plains this season constitutes around 228 days,whilst north of Gniezno and Szamotuł this slowly declines to 216 days.

Precipitation ranges from 500 to 550 mm. Despite this the region is still faced with a deficit in rainfall, particularly in the eastern part of the province (around Słupcy, Kazimierz Biskupi, Kleczew) where sometimes experience only 450 mm of rainfall per year, this threatens steppization of the region. Throughout the province there is typically a prevailing westerly wind.

Transportation

Greater Poland is a major transport hub within Poland, a great deal of traffic from Russia and other states of the former Soviet Union pass through Poznań and Konin to reach Germany and other EU member states. To the south runs the international route from Gdańsk via Poznań and Leszno to Prague and then to the south of Europe. There is also a major highway being built in in the province, the A2 motorway, which when completed will run from the western border of Poland with Germany, through Poznań to Warsaw and then onwards via Belarus to Moscow.

The A2 motorway traverses the voivodeship

The main railway hubs located in Greater Poland are Poznań, Piła and Ostrów Wielkopolski. PKP Intercity operate a number of trains a day between Warsaw and Berlin which provide a fast connection for the two cities also to Poznań. This route was the first in Poland, adapted for use by the European high-speed transportation system. In the near future the government expects to construct a high-speed rail line in the shape of a Y connecting Kalisz and Poznań from Łódź, Warsaw and Wrocław.

Poznań is the port of arrival for most international travellers as it is plays host to Ławica International Airport which has recently seen the second highest passenger growth rate in the country.

Politics

The Greater Poland voivodeship's government is headed by the province's voivode *(governor)* who is appointed by the Polish Prime Minister. The voivode is then assisted in performing his duties by the voivodeship's marshal, who is the appointed speaker for the voivodeship's executive and is elected by the sejmik *(provincial assembly)*. The current voivode of Greater Poland is Piotr Florek, whilst the present marshal is Marek Woźniak.

The Sejmik of Greater Poland consists of 39 members.

Current voivode of Greater Poland, Piotr Florek (front, 2nd from right), with marshal Marek Woźniak (front, 3rd from right)

Kuyavian-Pomeranian Regional Assembly elections on 21 November 2010[4]			
Party	**Votes**	**%**	**Total seats held**
Civic Platform (PO)	345,209	32.04	17
Law and Justice (PiS)	193,395	17.95	6
Left and Democrats (LiD)	232,704	21.60	9
Polish People's Party (PSL)	193,953	18.00	7
Others	112,171	10.41	0
Total	**1,077,356**	**100.00**	**39**

- Votes counted: 1,271,549
- Valid votes: 1,077,356
- **Turnout: 47.04%**

Administrative division

Greater Poland Voivodeship is divided into 35 counties (powiats): 4 city counties and 31 land counties. These are further divided into 226 gminas.

The counties are listed in the following table (ordering within categories is by decreasing population).

English and Polish names	Area		Population (2006)	Seat	Other towns	Total gminas
	(km²)	(sq mi)				
City counties						
Poznań	262	101	566,546			1
Kalisz	70	27	108,575			1
Konin	82	32	80,618			1
Leszno	32	12	64,079			1
Land counties						
Poznań County powiat poznański	1900	734	291,562	Poznań *	Swarzędz, Luboń, Mosina, Murowana Goślina, Puszczykowo, Kostrzyn, Pobiedziska, Kórnik, Buk, Stęszew	17
Ostrów Wielkopolski County powiat ostrowski	1161	448	158,407	Ostrów Wielkopolski	Nowe Skalmierzyce, Odolanów, Raszków	8
Gniezno County powiat gnieźnieński	1254	484	140,333	Gniezno	Witkowo, Trzemeszno, Kłecko, Czerniejewo	10
Piła County powiat pilski	1267	489	137,099	Piła	Wyrzysk, Ujście, Łobżenica, Wysoka	9
Konin County powiat koniński	1579	610	123,646	Konin *	Golina, Kleczew, Sompolno, Ślesin, Rychwał	14
Koło County powiat kolski	1011	390	88,601	Koło	Kłodawa, Dąbie, Przedecz	11
Czarnków-Trzcianka County powiat czarnkowsko-trzcianecki	1808	698	86,134	Czarnków	Trzcianka, Krzyż Wielkopolski, Wieleń	8
Szamotuły County powiat szamotulski	1120	432	85,849	Szamotuły	Wronki, Pniewy, Obrzycko, Ostroróg	8
Turek County powiat turecki	929	359	83,635	Turek	Tuliszków, Dobra	9
Kalisz County powiat kaliski	1160	448	80,369	Kalisz *	Stawiszyn	11
Kościan County powiat kościański	723	279	77,760	Kościan	Śmigiel, Czempiń, Krzywiń	5
Krotoszyn County powiat krotoszyński	714	276	77,092	Krotoszyn	Koźmin Wielkopolski, Zduny, Kobylin, Sulmierzyce	6
Gostyń County powiat gostyński	810	313	75,683	Gostyń	Krobia, Poniec, Borek Wielkopolski, Pogorzela	7
Września County powiat wrzesiński	704	272	73,778	Września	Miłosław, Nekla, Pyzdry	5

Nowy Tomyśl County powiat nowotomyski	1012	391	71,817	Nowy Tomyśl	Opalenica, Zbąszyń, Lwówek	6
Jarocin County powiat jarociński	588	227	70,390	Jarocin	Żerków	4
Złotów County powiat złotowski	1661	641	68,526	Złotów	Jastrowie, Okonek, Krajenka	8
Wągrowiec County powiat wągrowiecki	1041	402	67,606	Wągrowiec	Skoki, Gołańcz	7
Pleszew County powiat pleszewski	712	275	61,951	Pleszew		6
Rawicz County powiat rawicki	553	214	59,375	Rawicz	Miejska Górka, Bojanowo, Jutrosin	5
Słupca County powiat słupecki	838	324	58,725	Słupca	Zagórów	8
Śrem County powiat śremski	574	222	58,646	Śrem	Książ Wielkopolski, Dolsk	4
Oborniki County powiat obornicki	713	275	55,976	Oborniki	Rogoźno	3
Kępno County powiat kępiński	608	235	55,335	Kępno		7
Wolsztyn County powiat wolsztyński	680	263	54,718	Wolsztyn		3
Środa Wielkopolska County powiat średzki	623	241	54,568	Środa Wielkopolska		5
Ostrzeszów County powiat ostrzeszowski	772	298	54,490	Ostrzeszów	Grabów nad Prosną, Mikstat	7
Leszno County powiat leszczyński	805	311	50,024	Leszno *	Rydzyna, Osieczna	7
Grodzisk Wielkopolski County powiat grodziski	644	249	49,444	Grodzisk Wielkopolski	Rakoniewice, Wielichowo	5
Chodzież County powiat chodzieski	681	263	46,967	Chodzież	Szamocin, Margonin	5
Międzychód County powiat międzychodzki	737	285	36,329	Międzychód	Sieraków	4
* seat not part of the county						

Protected areas

Protected areas in Greater Poland Voivodeship include two National Parks and 12 Landscape Parks. These are listed below.

- Drawno National Park (partly in Lubusz and West Pomeranian Voivodeships)
- Wielkopolska National Park
- Barycz Valley Landscape Park (partly in Lower Silesian Voivodeship)
- Chłapowski Landscape Park
- Lednica Landscape Park
- Powidz Landscape Park
- Promno Landscape Park
- Przemęt Landscape Park (partly in Lubusz Voivodeship)
- Pszczew Landscape Park (partly in Lubusz Voivodeship)
- Puszcza Zielonka Landscape Park
- Rogalin Landscape Park
- Sieraków Landscape Park
- Warta Landscape Park
- Żerków-Czeszewo Landscape Park

Most popular surnames in the region

1. Nowak: 35,011
2. Kaczmarek: 24,185
3. Wojciechowski: 12,928

See also

- Prussia's Province of Posen (1818–1919)
- Second Polish Republic's Poznań Voivodeship (1921–1939)

References

[1] http://www.en.poznan.uw.gov.pl/

[2] http://www.nottinghamshire.gov.uk/home/your_council/howweprovideyourservices/partnerships/internationalandtwinning/transnationalpartnerships.htm

[3] http://www.stat.gov.pl/gus/45_655_PLK_HTML.htm

[4] "Greater Poland Regional Assembly elections" (http://wybory2010.pkw.gov.pl/geo/pl/300000/300000.html#tabs-5). State Electoral Commission. . Retrieved 2011-05-28.

Further reading

- Zygmunt Boras, *Książęta Piastowscy Wielkopolski* (Piast Princes of Wielkopolska), Poznań, Wydawnictwo Poznańskie, 1983.

External links

- wielkopolska-region.pl (http://www.wielkopolska-region.pl/) Information about Great Poland; Multilingual.
- greatpoland.eu (http://www.greatpoland.eu/) Data base about: Business, Culture, Sport, Motorization, Tourism, Agrotourism, Medicine, Health by CyberWielkopolska
- Greater Poland Local Government Office (http://www.wielkopolska.mw.gov.pl/)
- ChefMoz Dining Guide Greater Poland (http://chefmoz.org/Poland/WP/)
- Wielkopolska (http://www.dmoz.org/Regional/Europe/Poland/Voivodships/Wielkopolska/) at the Open Directory Project

Voivodeships_of_Poland

The **voivodeship**,[1] or **province**,[2] called in Polish ***województwo*** [vɔjɛˈvut͡stfɔ] (plural *województwa*), has been a high-level administrative subdivision of Poland since the 14th century.

The Polish local government reforms adopted in 1998, which went into effect on 1 January 1999, created sixteen new voivodeships. These replaced the 49 former voivodeships that had existed from 1 July 1975.

Today's voivodeships are mostly named after historical and geographical regions, while those prior to 1998 generally took their names from the cities on which they were centered. The new units range in area from under 10000 km^2 (3900 sq mi) (Opole Voivodeship) to over 35000 km^2 (14000 sq mi) (Masovian Voivodeship), and in population from one million (Lubusz Voivodeship) to over five million (Masovian Voivodeship).

Administrative authority at voivodeship level is shared between a government-appointed governor called a voivode (Polish *wojewoda*), an elected assembly called a sejmik, and an executive chosen by that assembly. The leader of that executive is called the *marszałek województwa* (voivodeship marshal). Voivodeships are further divided into powiats (counties) and gminas (communes or municipalities): see Administrative division of Poland.

Voivodeships since 1999

Administrative powers

Competences and powers at voivodeship level are shared between the voivode (governor), the sejmik (regional assembly) and the executive. In most cases these institutions are all based in one city, but in Kuyavian-Pomeranian and Lubusz Voivodeship the voivode's offices are in a different city from those of the executive and the sejmik. Voivodeship capitals are listed in the table below.

The **voivode** is appointed by the Prime Minister and is the regional representative of the central government. The voivode acts as the head of central government institutions at regional level (such as the police and fire services, passport offices, and various inspectorates), manages central government property in the region, oversees the functioning of local government, coordinates actions in the field of public safety and environment protection, and exercises special powers in emergencies. The voivode's offices collectively are known as the *urząd wojewódzki.*

The **sejmik** is elected every four years, at the same time as the local authorities at powiat and gmina level. It passes bylaws, including the voivodeship's development strategies and budget. It also elects the *marszałek* and other members of the executive, and holds them to account.

The **executive** (*zarząd województwa*), headed by the *marszałek*, drafts the budget and development strategies, implements the resolutions of the sejmik, manages the voivodeship's property, and deals with many aspects of regional policy, including management of European Union funding. Its offices collectively are known as the *urząd marszałkowski.*

Map and table of voivodeships

Polish voivodeships since 1999

Abbreviation	Coat of arms	code	car plates	Voivodeship	Capital	Area km²	Population (December 31, 2003)	Population (June 30, 2004)
DS		02	D	Lower Silesian (*dolnośląskie*)	Wrocław	19 947.76	2 898 313	2 895 729
KP		04	C	Kuyavian-Pomeranian (*kujawsko-pomorskie*)	Bydgoszcz[1] Toruń[2]	17 969.72	2 068 142	2 067 548
LU		06	L	Lublin (*lubelskie*)	Lublin	25 114.48	2 191 172	2 187 918
LB		08	F	Lubusz (*lubuskie*)	Gorzów Wielkopolski[1] Zielona Góra[2]	13 984.44	1 008 786	1 009 177
LD		10	E	Łódź (*łódzkie*)	Łódź	18 219.11	2 597 094	2 592 568
MP		12	K	Lesser Poland (*małopolskie*)	Kraków	15 144.10	3 252 949	3 256 171
MA		14	W	Masovian (*mazowieckie*)	Warsaw	35 597.80	5 135 732	5 139 545
OP		16	O	Opolskie	Opole	9 412.47	1 055 667	1 053 723
PK		18	R	Podkarpackie	Rzeszów	17 926.28	2 097 248	2 097 325
PD		20	B	Podlaskie (*podlaskie*)	Białystok	20 179.58	1 205 117	1 204 036
PM		22	G	Pomeranian (*pomorskie*)	Gdańsk	18 292.88	2 188 918	2 192 404
SL		24	S	Silesian (*śląskie*)	Katowice	12 294.04	4 714 982	4 707 825
SW		26	T	Świętokrzyskie	Kielce	11 672.34	1 291 598	1 290 176

WM		28	N	Warmian-Masurian (*warmińsko-mazurskie*)	Olsztyn	24 202.95	1 428 885	1 428 385
WP		30	P	Greater Poland (*wielkopolskie*)	Poznań	29 825.59	3 359 932	3 362 011
ZP		32	Z	West Pomeranian (*zachodniopomorskie*)	Szczecin	22 901.48	1 696 073	1 695 708
(1) – seat of voivode, (2) – seat of sejmik and marszałek								

See also:

- *Map of Polish Regions* [3]
- *Administrative division of Poland (from Commission on Standardization of Geographical Names Outside Poland website, in English)* [4]
- *Official map by Head Office of Geodesy and Cartography* [5]

Former voivodeships

Poland's voivodeships 1975–1998

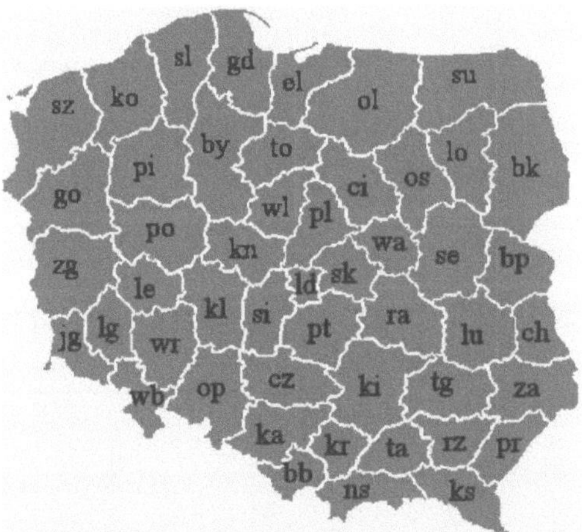

Administrative division of Poland between 1979 and 1998 included 49 voivodeships upheld after the establishment of the Third Polish Republic in 1989 for another decade. This reorganization of administrative division of Poland was mainly a result of local government reform acts of 1973–1975. In place of the three-level administrative division (voivodeship, county, commune), a new two-level administrative division was introduced (49 small voivodeships, and communes). The three smallest voivodeships – Warsaw, Kraków and Łódź – had the special status of municipal voivodeship; the city president (mayor) was also provincial governor.

Polish voivodeships and separate cities 1975-1998							
Abbreviation	**Voivodeship**	**Polish name**	**Capital**	**Area km^2 (1998)**	**Population (1980)**	**No. of cities**	**No. of communes**
bp	Biała Podlaska Voivodeship	bialskopodlaskie	Biała Podlaska	5 348	286 400	6	35
bk	Białystok Voivodeship	białostockie	Białystok	10 055	641 100	17	49
bb	Bielsko-Biała Voivodeship	bielskie	Bielsko-Biała	3 704	829 900	18	47
by	Bydgoszcz Voivodeship	bydgoskie	Bydgoszcz	10 349	1 036 000	27	55
ch	Chełm Voivodeship	chełmskie	Chełm	3 865	230 900	4	25
ci	Ciechanów Voivodeship	ciechanowskie	Ciechanów	6 362	405 400	9	45
cz	Częstochowa Voivodeship	częstochowskie	Częstochowa	6 182	747 900	17	49
el	Elbląg Voivodeship	elbląskie	Elbląg	6 103	441 500	15	37
gd	Gdańsk Voivodeship	gdańskie	Gdańsk	7 394	1 333 800	19	43
go	Gorzów Voivodeship	gorzowskie	Gorzów Wielkopolski	8 484	455 400	21	38
jg	Jelenia Góra Voivodeship	jeleniogórskie	Jelenia Góra	4 378	492 600	24	28
kl	Kalisz Voivodeship	kaliskie	Kalisz	6 512	668 000	20	53
ka	Katowice Voivodeship	katowickie	Katowice	6 650	3 733 900	43	46
ki	Kielce Voivodeship	kieleckie	Kielce	9 211	1 068 700	17	69
kn	Konin Voivodeship	konińskie	Konin	5 139	441 200	18	43
ko	Koszalin Voivodeship	koszalińskie	Koszalin	8 470	462 200	17	35
kr	Kraków Voivodeship	krakowskie	Kraków	3 254	1 167 500	10	38
ks	Krosno Voivodeship	krośnieńskie	Krosno	5 702	448 200	12	37
lg	Legnica Voivodeship	legnickie	Legnica	4 037	458 900	11	31
le	Leszno Voivodeship	leszczyńskie	Leszno	4 254	357 600	19	28
lu	Lublin Voivodeship	lubelskie	Lublin	6 793	935 200	16	62
lo	Łomża Voivodeship	łomżyńskie	Łomża	6 684	325 800	12	39
ld	Łódź Voivodeship	łódzkie	Łódź	1523	1 127 800	8	11
ns	Nowy Sącz Voivodeship	nowosądeckie	Nowy Sącz	5 576	628 800	14	41
ol	Olsztyn Voivodeship	olsztyńskie	Olsztyn	12 327	681 400	21	48
op	Opole Voivodeship	opolskie	Opole	8 535	975 000	29	61
os	Ostrołęka Voivodeship	ostrołęckie	Ostrołęka	6 498	371 400	9	38
pi	Piła Voivodeship	pilskie	Piła	8 205	437 100	24	35
pt	Piotrków Voivodeship	piotrkowskie	Piotrków Trybunalski	6 266	604 200	10	51
pl	Płock Voivodeship	płockie	Płock	5 117	496 100	9	44
po	Poznań Voivodeship	poznańskie	Poznań	8 151	1 237 800	33	57
pr	Przemyśl Voivodeship	przemyskie	Przemyśl	4 437	380 000	9	35
ra	Radom Voivodeship	radomskie	Radom	7 295	702 300	15	61
rz	Rzeszów Voivodeship	rzeszowskie	Rzeszów	4 397	648 900	13	41

se	Siedlce Voivodeship	siedleckie	Siedlce	8 499	616 300	12	66
si	Sieradz Voivodeship	sieradzkie	Sieradz	4 869	392 300	9	40
sk	Skierniewice Voivodeship	skierniewickie	Skierniewice	3 959	396 900	8	36
sl	Słupsk Voivodeship	słupskie	Słupsk	7 453	369 800	11	31
su	Suwałki Voivodeship	suwalskie	Suwałki	10 490	422 600	14	42
sz	Szczecin Voivodeship	szczecińskie	Szczecin	9 981	897 900	29	50
tg	Tarnobrzeg Voivodeship	tarnobrzeskie	Tarnobrzeg	6 283	556 300	14	46
ta	Tarnów Voivodeship	tarnowskie	Tarnów	4 151	607 000	9	41
to	Toruń Voivodeship	toruńskie	Toruń	5 348	610 800	13	41
wb	Wałbrzych Voivodeship	wałbrzyskie	Wałbrzych	4 168	716 100	31	30
wa	Warsaw Voivodeship	warszawskie	Warsaw	3 788	2 319 100	27	32
wl	Włocławek Voivodeship	włocławskie	Włocławek	4 402	413 400	14	30
wr	Wrocław Voivodeship	wrocławskie	Wrocław	6 287	1 076 200	16	33
za	Zamość Voivodeship	zamojskie	Zamość	6 980	472 100	5	47
zg	Zielona Góra Voivodeship	zielonogórskie	Zielona Góra	8 868	609 200	26	50

Poland's voivodeships 1945–1975

After World War II, the new administrative division of the country within the new national borders was based on the prewar one and included 14 (+2) voivodeships, then 17 (+5). The voivodeships in the east that had not been annexed by the Soviet Union had their borders left almost unchanged. The newly acquired territories in the west and north were organized into the new voivodeships of Szczecin, Wrocław and Olsztyn, and partly joined to Gdańsk, Katowice and Poznań voivodeships. Two cities were granted voivodeship status: Warsaw and Łódź.

In 1950, new voivodeships were created: Koszalin (previously part of Szczecin), Opole (previously part of Katowice), and Zielona Góra (previously part of Poznań, Wrocław and Szczecin voivodeships). In addition, three more cities were granted voivodeship status: Wrocław, Kraków and Poznań.

Polish administrative division 1945-1975				
Car plates (since 1956)	**Voivodeship**	**Capital**	**Area km² (1965)**	**Population (1965)**
A	białostockie	Białystok	23 136	1 160 400
B	bydgoskie	Bydgoszcz	20 794	1 837 100
G	gdańskie	Gdańsk	10 984	1 352 800
S	katowickie	Katowice	9 518	3 524 300
C	kieleckie	Kielce	19 498	1 899 100
E	koszalińskie [1]	Koszalin	17 974	755 100
K	krakowskie	Kraków	15 350	2 127 600
F	Łódzkie	Łódź	17 064	1 665 200
L	lubelskie	Lublin	24 829	1 900 500
O	olsztyńskie	Olsztyn	20 994	956 600
H	opolskie [1]	Opole	9 506	1 009 200
P	poznańskie	Poznań	26 723	2 126 300
R	rzeszowskie	Rzeszów	18 658	1 692 800
M	szczecińskie	Szczecin	12 677	847 600
T	warszawskie	Warsaw	29 369	2 453 000
X	wrocławskie	Wrocław	18 827	1 967 000
Z	zielonogórskie [1]	Zielona Góra	14 514	847 200
car plates (since 1956)	**Separate city**		**Area km² (1965)**	**Population (1965)**
I	Łódź		214	744 100
W	Warsaw		446	1 252 600
?	Kraków [2]		230	520 100
?	Poznań [2]		220	438 200
?	Wrocław [2]		225	474 200
[1] – new voivodeships created in 1950; [2] – cities separated in 1957				

Poland's voivodeships 1921–1939

Further information: Administrative division of Second Polish Republic

The administrative division of Poland in the interwar period included 16 voivodeships and Warsaw (with voivodeship rights).

Polish voivodeships in the interbellum (data as per April 1, 1937)					
car plates (since 1937)	Voivodeship Separate city	Capital	Area in 1000 km² (1930)	Population in 1000 (1931)	Map
00-19	City of Warsaw	Warsaw	0.14	1179.5	
85-89	warszawskie	Warsaw	31.7	2460.9	
20-24	białostockie	Białystok	26.0	1263.3	
25-29	kieleckie	Kielce	22.2	2671.0	
30-34	krakowskie	Kraków	17.6	2300.1	
35-39	lubelskie	Lublin	26.6	2116.2	
40-44	lwowskie	Lwów	28.4	3126.3	
45-49	łódzkie	Łódź	20.4	2650.1	
50-54	nowogródzkie	Nowogródek	23.0	1057.2	
55-59	poleskie	Brześć nad Bugiem	36.7	1132.2	
60-64	pomorskie	Toruń	25.7	1884.4	
65-69	poznańskie	Poznań	28.1	2339.6	
70-74	stanisławowskie	Stanisławów	16.9	1480.3	
75-79 ?	śląskie	Katowice	5.1	1533.5	
80-84	tarnopolskie	Tarnopol	16.5	1600.4	
90-94	wileńskie	Wilno	29.0	1276.0	
95-99	wołyńskie	Łuck	35.7	2085.6	

Congress Poland 1816–1837

Further information: Administrative division of Congress Poland

From 1816 to 1837 there were 8 voivodeships in Congress Poland.

- Augustów Voivodeship
- Kalisz Voivodeship
- Kraków Voivodeship
- Lublin Voivodeship
- Mazowsze Voivodeship
- Płock Voivodeship
- Podlasie Voivodeship
- Sandomierz Voivodeship

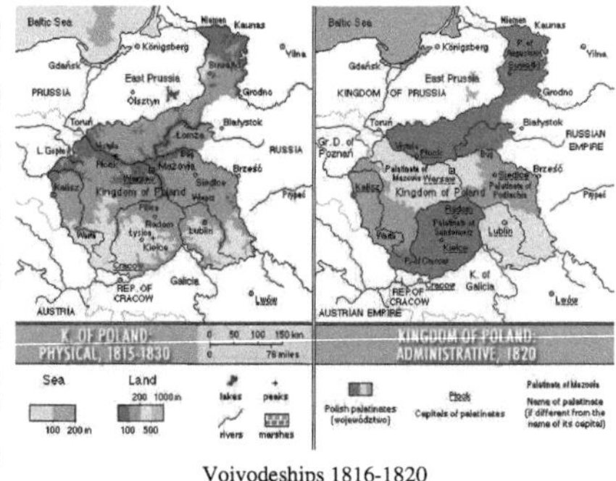

Voivodeships 1816-1820

Polish–Lithuanian Commonwealth 1569–1795

Further information: Administrative division of the Polish–Lithuanian Commonwealth and Voivodes of the Polish–Lithuanian Commonwealth

Greater Poland (*Wielkopolska*)

- Poznań Voivodeship (*województwo poznańskie*, Poznań)
- Kalisz Voivodeship (*województwo kaliskie*, Kalisz)
- Gniezno Voivodeship (*województwo gnieźnieńskie*, Gniezno) from 1768
- Sieradz Voivodeship (*województwo sieradzkie*, Sieradz)
- Łęczyca Voivodeship (*województwo łęczyckie*, Łęczyca)
- Brześć Kujawski Voivodeship (*województwo brzesko-kujawskie*, Brześć Kujawski)
- Inowrocław Voivodeship (*województwo inowrocławskie*, Inowrocław)
- Chełmno Voivodeship (*województwo chełmińskie*, Chełmno)
- Malbork Voivodeship (*województwo malborskie*, Malbork)
- Pomeranian Voivodeship (*województwo pomorskie*, Gdańsk)
- Duchy of Warmia (*Księstwo Warmińskie*, Lidzbark Warmiński)
- Duchy of Prussia (*Księstwo Pruskie*, Lidzbark Warmiński)
- Płock Voivodeship (*województwo płockie*, Płock)
- Rawa Voivodeship (*województwo rawskie*, Rawa Mazowiecka)
- Masovian Voivodeship (*województwo mazowieckie*, Warszawa)

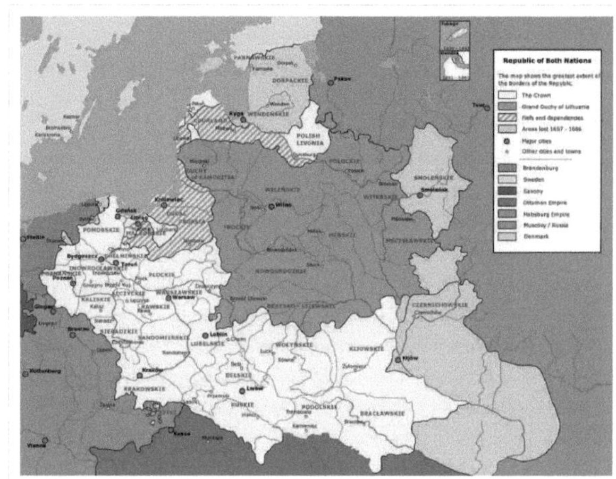

Voivodeships of the Republic of the Two Nations ("Polish–Lithuanian Commonwealth").

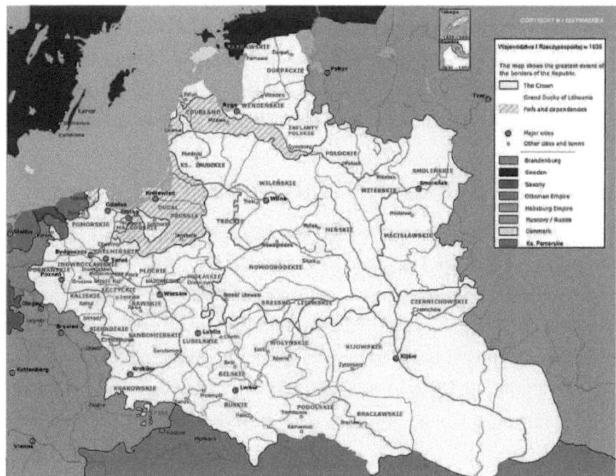

Voivodeships of the Commonwealth of the Two Nations in 1635

Lesser Poland (*Małopolska*)

- Kraków Voivodeship (*województwo krakowskie*, Kraków)
- Sandomierz Voivodeship (*województwo sandomierskie*, Sandomierz)
- Lublin Voivodeship (*województwo lubelskie*, Lublin)
- Podlaskie Voivodeship (*województwo podlaskie*, Drohiczyn)

- Ruthenian Voivodeship (*województwo ruskie*, Lwów)
- Bełz Voivodeship (*województwo bełzkie*, Bełz)
- Volhynian Voivodeship (*województwo wołyńskie*, Łuck)
- Podole Voivodeship (*województwo podolskie*, Kamieniec Podolski)
- Bracław Voivodeship (*województwo bracławskie*, Bracław)
- Kijów Voivodeship (*województwo kijowskie*, Kijów)
- Czernihów Voivodeship (*województwo czernichowskie*, Czernihów)

Grand Duchy of Lithuania

Here the first name given is English, then in brackets – Lithuanian, and then Polish.

- Wilno Voivodeship (*Vilniaus vaivadija, województwo wileńskie*, Vilnius)
- Troki Voivodeship (*Trakų vaivadija, województwo trockie*, Trakai)
- Nowogródek Voivodeship (*Naugarduko vaivadija, województwo nowogrodzkie*, Nowogródek)
- Brest-Litovsk Voivodeship (*Lietuvos Brastos vaivadija, województwo brzesko-litewskie*, Brześć Litewski)
- Minsk Voivodeship (*Minsko vaivadija, województwo mińskie*, Mińsk)
- Mścisław Voivodeship (*Mstslavlio vaivadija, województwo mścisławskie*, Mscislaw)
- Smolensk Voivodeship (*Smolensko vaivadija, województwo smoleńskie*, Smoleńsk)
- Vitebsk Voivodeship (*Vitebsko vaivadija, województwo witebskie*, Witebsk)
- Połock Voivodeship (*Polocko vaivadija, województwo połockie*, Połock)
- Duchy of Samogita (*Žemaičių seniūnija, księstwo żmudzkie*, Medininkai-Varniai)

Duchy of Livonia

- Wenden Voivodeship (*województwo wendeńskie*, Wenden) since 1598 till the 1620s
- Dorpat Voivodeship (*województwo dorpackie*, Dorpat) since 1598 till the 1620s
- Parnawa Voivodeship (*województwo parnawskie*, Parnawa) since 1598 till the 1620s
- Inflanty Voivodeship (*województwo inflanckie*, Dyneburg) since the 1620s
- Duchy of Courland and Semigalia (*księstwo Kurlandii i Semigalii*, Mitawa)

Etymology and use of "voivodeship"

Some English-language sources, in historic contexts, speak of "palatinates" rather than "voivodeships"; the former term traces back to the Latin *palatinus* ("palatine"). More commonly used now is "voivodeship", a loanword-calque hybrid formed on the Polish "*województwo*". Other sources refer instead to "provinces" (Polish singular: "*prowincja*"), though in pre-1795 contexts this may be confusing because the cognate Polish "*prowincyja*" (as it was then spelled) was idiosyncratically applied, until the last of the three Partitions of the Polish–Lithuanian Commonwealth, in 1795, to each of the three main Regions (Greater Poland, Lesser Poland, and Lithuania) of the Polish–Lithuanian Commonwealth, each of those Regions in turn comprising a number of *województwa* (plural of "*województwo*").

The Polish "*województwo*", designating a second-tier Polish or Polish–Lithuanian administrative unit, derives from "*wojewoda*" (etymologically, a "war leader" or "leader of warriors", but now simply the governor of a *województwo*) and the suffix "*-stwo*" (a "state or condition").

The English "voivodeship", which is a hybrid of the loanword "voivode" and "-ship" (the latter a suffix, likewise meaning a "state or condition", that calques the Polish "*-stwo*"), has never been much used and is absent from many dictionaries. According to the *Oxford English Dictionary*, it first appeared in 1792, spelled "woiwodship", in the sense of "the district or province governed by a voivode." The word subsequently also appeared in 1886 in the sense of "the office or dignity of a voivode."[6]

An official Polish body, the Commission on Standardization of Geographic Names outside the Republic of Poland, recommends the spelling "voivodship", without the *e*. This is consistently reflected in publications and in the international arena, e.g., at the United Nations.

See also

- Voivodeships of Poland (1975–1998)
- Administrative division of the Polish–Lithuanian Commonwealth
- Coats of arms of Polish voivodeships
- Flags of Polish voivodeships
- ISO 3166-2:PL
- *Prowincja*
- Regions of Poland
- Voivodeship

Notes

[1] Also spelled "voivodship," "voi**e**vodship," "voi**e**vod**e**ship".

[2] The word "voivodeship" appears in some larger English dictionaries, such as the OED and Webster's Third New International Dictionary, but it is not in common usage. Thus, to facilitate understanding outside Poland, the word "province" is a recommended translation: "*Jednostki podziału administracyjnego Polski tłumaczymy tak: województwo*—province..." ("Polish administrative units are translated as follows: *województwo*—province..."). Arkadiusz Belczyk, *"Tłumaczenie polskich nazw geograficznych na język angielski"* (http://serwistlumacza.com/content/view/27/32/) ("Translation of Polish Geographical Names into English"), 2002-2006. Examples: New Provinces of Poland (1998) (http://www.staff.amu.edu.pl/~zbzw/ph/pro/plpro.html), Map of Poland (http://www.map-of-Poland.co.uk/), English names of Polish provinces (http://serwistlumacza.com/content/view/78/). More examples:

- "Following the reform of the administrative structure in 1973-1975, the number of **provinces** (*województwa*) was increased from 22 to 49... [I]ncreasing the number of **provinces** meant the reduction of each in size. In this way Warsaw was able to dilute the political importance of the **provincial** party chiefs." "Poland", *The Encyclopedia Americana*, 1986, volume 22, p. 312.
- "Poland is divided into 49 **provinces**." "Poland", *The Columbia Encyclopedia*, sixth edition, edited by Paul Lagassé, Columbia University Press, 2000, p. 2256.
- "Local government in Poland is organized on three levels. The largest units, at the regional level, are the *województwa* (**provinces**)..." "Poland", *Encyclopaedia Britannica*, 15th edition, 2010, *Macropaedia*, volume 25, p. 937.
- "**GOVERNMENT**... **Administrative divisions**: 16 **provinces** (wojewodztwa, singular–wojewodztwo)..." "Poland," in Central Intelligence Agency, *The CIA World Factbook 2010*, New York, Skyhorse Publishing, Inc., 2009, ISBN 978-60239-727-9, p. 546. The same information appears in the current online *CIA World Factbook* --> "Poland --> Administrative divisions". (https://www.cia.gov/library/publications/the-world-factbook/geos/pl.html) Note that in this source, where "English translations" of province names are given, they are in the noun ("Silesia"), not the adjective ("Silesia**n**"), form.

[3] http://www.rzeczpospolita.pl/teksty/wydanie_040302/kraj_a_7-1.F.jpg

[4] http://www.gugik.gov.pl/komisja/english/pliki/podzial_administracyjny_polski_2006-eng.pdf

[5] http://www.gugik.gov.pl/komisja/english/pliki/polska_2006-podzial_administracyjny.jpg

[6] "Voivodeship," *The Oxford English Dictionary*, second edition, volume XIX, Oxford, Clarendon Press, 1989, p. 739.

References

- "Poland", *Encyclopaedia Britannica*, 15th edition, 2010, *Macropaedia*, volume 25, p. 937.
- "Poland", *The Columbia Encyclopedia*, sixth edition, edited by Paul Lagassé, Columbia University Press, 2000, p. 2256.
- "Poland", *The Encyclopedia Americana*, 1986, volume 22, p. 312.
- "Poland," in Central Intelligence Agency, *The CIA World Factbook 2010*, New York, Skyhorse Publishing, Inc., 2009, ISBN 978-60239-727-9, p. 546.
- "Voivodeship," *The Oxford English Dictionary*, second edition, volume XIX, Oxford, Clarendon Press, 1989, p. 739.

External links

- Regions of Poland (http://regiony.poland.gov.pl/start_en.html)
- Toponymic Guidelines Of Poland for Map Editors and Other Users (http://ksng.gugik.gov.pl/pliki/topon_inte.doc.pdf) Head Office Of Geodesy And Cartography, 2002
- *CIA World Factbook* --> "Poland --> Administrative divisions" (https://www.cia.gov/library/publications/the-world-factbook/geos/pl.html)

Borówiec

Borówiec	
— Village —	
 Borówiec	
Coordinates: 52°17′N 17°2′E	
Country	Poland
Voivodeship	Greater Poland
County	Poznań County
Gmina	Kórnik
Population	1603

Borówiec [bɔˈruvjɛt͡s] is a village in the administrative district of Gmina Kórnik, within Poznań County, Greater Poland Voivodeship, in west-central Poland.[1] It lies approximately 7 kilometres (4 mi) north-west of Kórnik and 16 km (10 mi) south-east of the regional capital Poznań.

The village has a population of 1,603.

References

[1] "Central Statistical Office (GUS) - TERYT (National Register of Territorial Land Apportionment Journal)" (http://www.stat.gov.pl/broker/access/prefile/listPreFiles.jspa) (in Polish). 2008-06-01. .

Article Sources and Contributors

Drogosławiec *Source*: http://en.wikipedia.org/w/index.php?title=Drogos%C5%82awiec *Contributors*:

Bylin *Source*: http://en.wikipedia.org/w/index.php?title=Bylin *Contributors*:

Village *Source*: http://en.wikipedia.org/w/index.php?title=Village *Contributors*: 1223Sallybride, 21655, A.Hassen, AKeen, Aarohan Mettu, Abductive, AdamDeanHall, Adamhauner, Airjatt, Aitias, Alaexis, Ale jrb, AlefZet, Aleksandr Grigoryev, Alensha, Altenmann, Amberrock, Aminullah, Andrewpmk, Andycjp, Animum, Apalamarchuk, Apokrif, Arpingstone, Arthena, AussieLegend, Azreey, Bejinhan, Beland, Belirac, Berig, Big Bird, Bill37212, Biruitorul, Bjh21, Bkonrad, Bobrayner, BocoROTH, Bogdan, Bogdangiusca, Bomac, Branka France, BreakingFree40, Brian the Editor, Brian0918, Butko, C.Fred, Caiaffa, CalJW, CalumMcA, Canley, Capricorn42, Cassowary, Catgut, Catslisten, Cecropia, Cenarium, Ceriy, Cgoodwin, Chasnor15, Chentchen, ChiragPatnaik, Chris j wood, Christopher.e.dunn, Chzz, Ciudad jardin, Cntras, Colonies Chris, Conleyb15, Conorbrady.ie, Corregere, Cosmic Latte, CrimsonBlue, Cwolfsheep, D6, DITWIN GRIM, Da-drumer, DailyWikiHelp, Daniel Case, Daniel Quinlan, David Kernow, Dbg1233, Ddstretch, Deathawk, Deeptrivia, Den fjättrade ankan, Deqon, Deror avi, Desiphral, Dezaree, Dieter Simon, Discospinster, Doctor Whom, Donarreiskoffer, Doseiai2, Dottod, Dpkipling, Dysepsion, E ponti, E0steven, Ehrenkater, Elassint, Eleassar, Ellywa, Elockid, Elroygoh02, Emyr93, Epipelagic, Erianna, Eu.stefan, EugeneZelenko, EurekaLott, Extra999, Ezhiki, Fang 23, Farosdaughter, Fat pig73, Flewis, Francis Tyers, Francs2000, Fred Bauder, Frosted14, Fujurcitook, Funandtrvl, FunkyFly, Futurebird, Gardar Rurak, Garhowell, Gegnome, Gfoley4, Gingermint, Ginkgo100, Giro720, Glenn, Globalphilosophy, GoOdCoNtEnT, Gogo Dodo, Gorai, Grafikm fr, GrahamSmithe, Gtswiki, Guanaco, Guss2, Gustavo Szwedowski de Korwin, Gz33, Hahnchen, Halsteadk, Harisu1988, Hayabusa future, Hebrides, Helenmwarwick, HenryLi, HighSpeed-X, Hmains, Hroðulf, Hu12, Hydrogen Iodide, IGeMiNix, IShadowed, IW.HG, Ianisbored, Indiaforever, Ineffable3000, Isam, IstvanWolf, IvanKlinko, J.delanoy, JHunterJ, JPSheridan, Jahangard, James086, Janetleopold, Jared Preston, Jcuk, JdeJ, Jeremy Bolwell, Jerrch, Jno, Jongleur100, Julesd, Julia W, JuneGloom07, KR3ColinB, Kajasudhakarababu, Kamiekam, Keelanrush, Kelisi, Kelovy, Khalid Mahmood, Khanvez, Kikos, Kimberry352, Kironbd07, KirrVlad, Kjramesh, Kman543210, Kmsasidhar, Komalrajiana, Ktr101, Kurruption, Kwamikagami, Larucio, Laurinavicius, Ldonna, Leonig Mig, Levineps, Lew121, LiDaobing, Lightmouse, LilHelpa, Linaduliban, Little muddy funkster, Littlealien182, Llywrch, Lonewolf BC, Lozleader, Luk, Lulu1127, Lumos3, Lynda Finn, MacTire02, Maglame, Maitch, Majorly, Malick78, Mallakottai, Mani1, Manly 2008, Marek69, Mariano12 1989, Mariwan zanko, Markmark12, Materialscientist, Mayooranathan, Mdzafri, Meldor, Mfo7787, MickMacNee, MisfitToys, MisterVodka, Modulatum, Moebiusuibeom-en, Moonriddengirl, Mrmuk, Mtaylor848, Mvjs, Mwalcoff, Mxn, NE2, Naturenet, Naveen Sankar, NerdyScienceDude, Nethergreen55, NewEnglandYankee, Nickshanks, Nistra, Noclevername, Nricardo, Nyttend, O1ive, Old Moonraker, Orangemike, OscarKosy, P.Marlow, Parkwells, Patrick, PaulBannister, Penrithguy, Peter E. James, PeterCScott, Pgan002, Phagopsych, Pharaoh of the Wizards, PhilKnight, Pinethicket, PleaseStand, Pokrajac, Ppeace, Pug1776, Puneet 348, Quaysys, R'n'B, RDF, Raeky, Randhirreddy, Rcingham, Realteacher, Reeses14, Reinyday, Rich Farmbrough, Rickyrab, Risker, Rjm at sleepers, Robdav69, Rocket71048576, Rwxrwxrwx, Saccerzd, Sadrettin, Saga City, Same7same7, SamuraiClinton, Samyo, Sansonic, Sardanaphalus, Schzmo, Seb az86556, Sem10efg, Sephiroth BCR, Sfu, Shantavira, Sheddybey, Shell Kinney, Shoeofdeath, SilkTork, Sjakkalle, Sketchmoose, Sole Soul, StephenDawson, Steven Zhang, Stevo D, Stifle, Student BSMU, Suffusion of Yellow, Suprgye, Surya Prakash.S.A., Szambo, THATSBETTER, Taamu, Tbhotch, Teinesavaii, TenIslands, Tenth Plague, The Thing That Should Not Be, The Transhumanist, Theranos, Thomasrflagel, Tide rolls, Tjanke, Tobby72, Tobias Conradi, Tpbradbury, Triona, Tsuchiya Hikaru, TurkChan, Twanner921, Twin Bird, Tytyty456, Ukabia, UnicornTapestry, Untifler, Utcursch, V2k, VMS Mosaic, Vegaswikian, Vhdwikiuser, Vikramdidawat, Vmenkov, WOSlinker, Warofdreams, Wayland, Wayne Slam, Wendy1991, West Brom 4ever, Whytecypress, Wiki Wikardo, Wikijens, William Allen Simpson, William Avery, Winhunter, Wknight94, WojPob, Woohookitty, WorldWide Update, Wrelwser43, Xiaoyu of Yuxi, Xyz or die, Yakudza, Yonkie, Yupik, Zacharie Grossen, Zhou Yu, Zoeyzazu, Zollerriia, Zotel, Александър, პაატა შ, , , , 419 anonymous edits

Gmina_Kórnik *Source*: http://en.wikipedia.org/w/index.php?title=Gmina_K%C3%B3rnik *Contributors*:

Poznań_County *Source*: http://en.wikipedia.org/w/index.php?title=Pozna%C5%84_County *Contributors*: A bit iffy, Ahadley, Ahoerstemeier, Balcer, Emax, Everyking, Halibutt, Komorniki, Kotniski, Logologist, Nricardo, PRB, PolishPoliticians, Radomil, Rigadoun, SFGiants, XRiffRaffx, 5 anonymous edits

Greater_Poland_Voivodeship *Source*: http://en.wikipedia.org/w/index.php?title=Greater_Poland_Voivodeship *Contributors*: A bit iffy, Ajh1492, Al Silonov, Amcde, Appleseed, Balcer, Barticus88, Betacommand, Biruitorul, Bonás, Bushy moustache, Bwood, Caius2ga, Chanheigeorge, CharlotteWebb, Chris G, Colonies Chris, CommonsDelinker, Cryptic, David Parker, Dominus Vobisdu, El., Elapsed, Elonka, Emax, Flatterworld, Gnesener1900, Greenshed, GvmBG, Halibutt, Hallihallo61, Joy, Julo, Kotniski, Kpjas, LilHelpa, LittleDan, Logologist, Lysy, MJCdetroit, MaGioZal, Marek69, Mazur, Mix321, NaveedLLB, Nihil novi, Odder, Ohconfucius, Orioane, PerV, Piotrus, Plastikspork, PolishPoliticians, ProhibitOnions, Radomil, Sardanaphalus, Scipius, Slawojarek, Socialservice, Spacejam2, Sylwia Ufnalska, Szczecin, Template namespace initialisation script, Temuri rajavi, Tim!, Topory, Umb is gay, WLKP, Wik, William Allen Simpson, Windharp, Youandme, Zbihniew, Zoe, Јованвб, 41 anonymous edits

Voivodeships_of_Poland *Source*: http://en.wikipedia.org/w/index.php?title=Voivodeships_of_Poland *Contributors*: Abhijitsathe, Adz, Ajh1492, Akerans, Alan Liefting, Alpha Quadrant, Altenmann, Andres, Aotearoa, Appleseed, B7628, Barticus88, BionicWilliam, Biruitorul, Bjung, Bonás, Brisvegas, Brockert, Byrial, CZmarlin, Caius2ga, Cautious, Chanheigeorge, Christianna1219, Colonies Chris, CrniBombarder!!!, Crusoe8181, DO'Neil, Danim, David Kernow, David Parker, DeirYassin, Deville, Docu, Domino theory, Dominus Vobisdu, Electionworld, Elonka, Emaus, Emax, Erwin, Fayenatic london, Gauss, Gnesener1900, Halibutt, Hmains, JDavid, JNW, Jauhienij, Jopxton, Juzeris, Kelisi, Kotniski, Kpalion, Kpjas, Kwamikagami, LinasLit, Logologist, Lysy, Marcin Suwalczan, Marek69, Mathiasrex, Maxí, MeltBanana, Meursault2004, Miranche, Naive cynic, Nihil novi, NoychoH, Nuno Tavares, Olessi, Orioane, OwenBlacker, Pak21, Patrick, Phil Boswell, Piotr Osada, Piotrus, Polish29, Poznaniak, Rarelibra, Reaper Eternal, RedWolf, Rjwilmsi, Roke, Sangjinhwa, Sanya, Sardanaphalus, Scipius, Severo, Skäpperöd, Slawojarek, Spacejam2, Steff, Sterlingsilver850, SwPawel, Sylwia Ufnalska, Tar-ba-gan, Template namespace initialisation script, Topbanana, Ttwaring, Uncle Milty, Valentinian, Vanjagenije, Volunteer Marek, WLKP, Wik, Wmahan, Worobiew, Xx236, 56 anonymous edits

Borówiec *Source*: http://en.wikipedia.org/w/index.php?title=Bor%C3%B3wiec *Contributors*:

Image Sources, Licenses and Contributors

File:Flag of Poland.svg *Source*: http://en.wikipedia.org/w/index.php?title=File:Flag_of_Poland.svg *License*: unknown *Contributors*: Anomie, Mifter

file:Poland location map.svg *Source*: http://en.wikipedia.org/w/index.php?title=File:Poland_location_map.svg *License*: unknown *Contributors*: User:NordNordWest

File:Red pog.svg *Source*: http://en.wikipedia.org/w/index.php?title=File:Red_pog.svg *License*: unknown *Contributors*: Anomie

File:Dartlo (2).jpg *Source*: http://en.wikipedia.org/w/index.php?title=File:Dartlo_(2).jpg *License*: unknown *Contributors*: Lidia Ilona

File:Ourika berbere village.jpg *Source*: http://en.wikipedia.org/w/index.php?title=File:Ourika_berbere_village.jpg *License*: unknown *Contributors*: Fathzer / Jean-Marc Astesana from Bretonneux, France

File:Rougon Alpes de Haute Provence France.jpg *Source*: http://en.wikipedia.org/w/index.php?title=File:Rougon_Alpes_de_Haute_Provence_France.jpg *License*: unknown *Contributors*: User:JialiangGao

File:kasipur2.jpg *Source*: http://en.wikipedia.org/w/index.php?title=File:Kasipur2.jpg *License*: unknown *Contributors*: Kironbd07

File:masouleh.jpg *Source*: http://en.wikipedia.org/w/index.php?title=File:Masouleh.jpg *License*: unknown *Contributors*: Original uploader was Hoomanb at en.wikipedia

File:KippelLötschental WoodenHouses.jpg *Source*: http://en.wikipedia.org/w/index.php?title=File:KippelLötschental_WoodenHouses.jpg *License*: unknown *Contributors*: Herzi Pinki

File:Cumalıkızık 7121.jpg *Source*: http://en.wikipedia.org/w/index.php?title=File:Cumalıkızık_7121.jpg *License*: unknown *Contributors*: User:Darwinek

File:Indianvillage.jpg *Source*: http://en.wikipedia.org/w/index.php?title=File:Indianvillage.jpg *License*: unknown *Contributors*: AKA MBG, Deeptrivia, Extra999, Roland zh

File:Ogi Shirakawa02bs3200.jpg *Source*: http://en.wikipedia.org/w/index.php?title=File:Ogi_Shirakawa02bs3200.jpg *License*: unknown *Contributors*: 663highland

File:Pariangan.jpg *Source*: http://en.wikipedia.org/w/index.php?title=File:Pariangan.jpg *License*: unknown *Contributors*: User:MichaelJLowe

File:Kampung house in Sungai Nipah.jpg *Source*: http://en.wikipedia.org/w/index.php?title=File:Kampung_house_in_Sungai_Nipah.jpg *License*: unknown *Contributors*: Bin Gregory

File:Kovachevitsa.jpg *Source*: http://en.wikipedia.org/w/index.php?title=File:Kovachevitsa.jpg *License*: unknown *Contributors*: User:Katieleu

File:Zaponorye 9173.jpg *Source*: http://en.wikipedia.org/w/index.php?title=File:Zaponorye_9173.jpg *License*: unknown *Contributors*: User:Gastro-en

File:castle.combe.mainstreet.arp.jpg *Source*: http://en.wikipedia.org/w/index.php?title=File:Castle.combe.mainstreet.arp.jpg *License*: unknown *Contributors*: Arpingstone, Troxx

File:Bisley.jpg *Source*: http://en.wikipedia.org/w/index.php?title=File:Bisley.jpg *License*: unknown *Contributors*: User:Jongleur100

File:Saint-Cirq-Lapopie.jpg *Source*: http://en.wikipedia.org/w/index.php?title=File:Saint-Cirq-Lapopie.jpg *License*: unknown *Contributors*: Adam Baker from Houston / Moscow / Toulouse (travel a lot)

File:saifivillage.JPG *Source*: http://en.wikipedia.org/w/index.php?title=File:Saifivillage.JPG *License*: unknown *Contributors*: Elie plus, Linaduliban

File:General view to Al annaze1.jpg *Source*: http://en.wikipedia.org/w/index.php?title=File:General_view_to_Al_annaze1.jpg *License*: unknown *Contributors*: User:Same7same7

File:Puamau.jpg *Source*: http://en.wikipedia.org/w/index.php?title=File:Puamau.jpg *License*: unknown *Contributors*: American at de.wikipedia

File:Carlb-fogo-newfoundland-2002.jpg *Source*: http://en.wikipedia.org/w/index.php?title=File:Carlb-fogo-newfoundland-2002.jpg *License*: unknown *Contributors*: Conscious, Croquant, EvaK, Peregrine981, Telim tor, WayneRay, 1 anonymous edits

File:POL powiat poznański flag.svg *Source*: http://en.wikipedia.org/w/index.php?title=File:POL_powiat_poznański_flag.svg *License*: unknown *Contributors*: user:Poznaniak

File:POL powiat poznański COA.svg *Source*: http://en.wikipedia.org/w/index.php?title=File:POL_powiat_poznański_COA.svg *License*: unknown *Contributors*: user:Poznaniak

File:POL województwo wielkopolskie powiat poznański map.svg *Source*: http://en.wikipedia.org/w/index.php?title=File:POL_województwo_wielkopolskie_powiat_poznański_map.svg *License*: unknown *Contributors*: user:WarX

File:POL województwo wielkopolskie flag.svg *Source*: http://en.wikipedia.org/w/index.php?title=File:POL_województwo_wielkopolskie_flag.svg *License*: unknown *Contributors*: Bastianow, Frombenny, Gustavo Szwedowski de Korwin, Ninane, Yarl, 2 anonymous edits

File:Powiat poznanski-map.png *Source*: http://en.wikipedia.org/w/index.php?title=File:Powiat_poznanski-map.png *License*: unknown *Contributors*: Original uploader was Roweromaniak (Stanisław Nowak) at pl.wikipedia. Later version(s) were uploaded by Masur at pl.wikipedia.

File:Logo Wielkopolskie.png *Source*: http://en.wikipedia.org/w/index.php?title=File:Logo_Wielkopolskie.png *License*: unknown *Contributors*: Bushy moustache

File:POL województwo wielkopolskie COA.svg *Source*: http://en.wikipedia.org/w/index.php?title=File:POL_województwo_wielkopolskie_COA.svg *License*: unknown *Contributors*: User:Bastianow

File:Wielkopolskie (EE,E NN,N).png *Source*: http://en.wikipedia.org/w/index.php?title=File:Wielkopolskie_(EE,E_NN,N).png *License*: unknown *Contributors*: user:Wulfstan, user:tsca

File:Woj wielkopolskie adm.png *Source*: http://en.wikipedia.org/w/index.php?title=File:Woj_wielkopolskie_adm.png *License*: unknown *Contributors*: MariuszR, Mir, Niki K, Yarl

File:Speaker Icon.svg *Source*: http://en.wikipedia.org/w/index.php?title=File:Speaker_Icon.svg *License*: unknown *Contributors*: Blast, G.Hagedorn, Mobius, 2 anonymous edits

File:Kórnik Zamek 234-07.jpg *Source*: http://en.wikipedia.org/w/index.php?title=File:Kórnik_Zamek_234-07.jpg *License*: unknown *Contributors*: Madmedea, Pko, Roweromaniak, Sir Gawain, 1 anonymous edits

File:Powstancy Wlkp.jpg *Source*: http://en.wikipedia.org/w/index.php?title=File:Powstancy_Wlkp.jpg *License*: unknown *Contributors*: Masur, Pascal.Tesson, Radomil, Topory

Image:Poznan.StaryRynek.jpg *Source*: http://en.wikipedia.org/w/index.php?title=File:Poznan.StaryRynek.jpg *License*: unknown *Contributors*: User:Łukasz Poznań

File:PL Kalisz Ratusz..JPG *Source*: http://en.wikipedia.org/w/index.php?title=File:PL_Kalisz_Ratusz..JPG *License*: unknown *Contributors*: Bast, Fransvannes, 3 anonymous edits

Image:Gniezno dom.JPG *Source*: http://en.wikipedia.org/w/index.php?title=File:Gniezno_dom.JPG *License*: unknown *Contributors*: User:Südstädter

File:Jezioro Góreckie.jpg *Source*: http://en.wikipedia.org/w/index.php?title=File:Jezioro_Góreckie.jpg *License*: unknown *Contributors*: User:Awersowy

Image:Power plant in Pątnów.jpg *Source*: http://en.wikipedia.org/w/index.php?title=File:Power_plant_in_Pątnów.jpg *License*: unknown *Contributors*: Tomasz Krzykała

Image:Autostrada A2 - Koło.jpg *Source*: http://en.wikipedia.org/w/index.php?title=File:Autostrada_A2_-_Koło.jpg *License*: unknown *Contributors*: w:pl:user:KolaninKolanin

File:Polish presidential election 2010, Poznań (10) - Woźniak, Florek.JPG *Source*: http://en.wikipedia.org/w/index.php?title=File:Polish_presidential_election_2010,_Poznań_(10)_-_Woźniak,_Florek.JPG *License*: unknown *Contributors*: User:Yves6

File:Poland administrative division 1999 literki.png *Source*: http://en.wikipedia.org/w/index.php?title=File:Poland_administrative_division_1999_literki.png *License*: unknown *Contributors*: Closedmouth, Dantadd, LMK3, Vladek Komorek

File:POL woj dolnoslaskie COA 2009.svg *Source*: http://en.wikipedia.org/w/index.php?title=File:POL_woj_dolnoslaskie_COA_2009.svg *License*: unknown *Contributors*: User:Mboro

File:POL województwo kujawsko-pomorskie COA.svg *Source*: http://en.wikipedia.org/w/index.php?title=File:POL_województwo_kujawsko-pomorskie_COA.svg *License*: unknown *Contributors*: Lech Tadeusz Karczewski

File:POL województwo lubelskie COA.svg *Source*: http://en.wikipedia.org/w/index.php?title=File:POL_województwo_lubelskie_COA.svg *License*: unknown *Contributors*: user:MesserWoland

File:POL województwo lubuskie COA.svg *Source*: http://en.wikipedia.org/w/index.php?title=File:POL_województwo_lubuskie_COA.svg *License*: unknown *Contributors*: User:Bastianow

File:POL województwo łódzkie COA.svg *Source*: http://en.wikipedia.org/w/index.php?title=File:POL_województwo_łódzkie_COA.svg *License*: unknown *Contributors*: User:Bastianow

File:POL województwo małopolskie COA.svg *Source*: http://en.wikipedia.org/w/index.php?title=File:POL_województwo_małopolskie_COA.svg *License*: unknown *Contributors*: user:Poznaniak

File:POL województwo mazowieckie COA.svg *Source*: http://en.wikipedia.org/w/index.php?title=File:POL_województwo_mazowieckie_COA.svg *License*: unknown *Contributors*: User:Poznaniak

File:POL województwo opolskie COA.svg *Source*: http://en.wikipedia.org/w/index.php?title=File:POL_województwo_opolskie_COA.svg *License*: unknown *Contributors*: user:Poznaniak

File:POL województwo podkarpackie COA.svg *Source*: http://en.wikipedia.org/w/index.php?title=File:POL_województwo_podkarpackie_COA.svg *License*: unknown *Contributors*: Derbeth, Frombenny, Gustavo Szwedowski de Korwin, Jaron11, Mir, Valentinian

File:POL województwo podlaskie COA.svg *Source*: http://en.wikipedia.org/w/index.php?title=File:POL_województwo_podlaskie_COA.svg *License*: unknown *Contributors*: Bastianow, Derbeth, Frombenny, Gustavo Szwedowski de Korwin, JDavid, WarX, XtraVert, 2 anonymous edits

File:POL województwo pomorskie COA.svg *Source*: http://en.wikipedia.org/w/index.php?title=File:POL_województwo_pomorskie_COA.svg *License*: unknown *Contributors*: User:Bastianow

File:POL województwo śląskie COA incorrect.svg *Source*: http://en.wikipedia.org/w/index.php?title=File:POL_województwo_śląskie_COA_incorrect.svg *License*: unknown *Contributors*: user:MesserWoland

File:POL wojewodztwo świętokrzyskie COA.svg *Source*: http://en.wikipedia.org/w/index.php?title=File:POL_wojewodztwo_świętokrzyskie_COA.svg *License*: unknown *Contributors*: User:Bastianow

File:Warminsko-mazurskie herb.svg *Source*: http://en.wikipedia.org/w/index.php?title=File:Warminsko-mazurskie_herb.svg *License*: unknown *Contributors*: Derbeth, Frombenny, Gustavo Szwedowski de Korwin, JDavid, MAC13, Serdelll, 1 anonymous edits

File:POL województwo zachodniopomorskie COA.svg *Source*: http://en.wikipedia.org/w/index.php?title=File:POL_województwo_zachodniopomorskie_COA.svg *License*: unknown *Contributors*: User:Bastianow

File:Poland administrative division 1975 literki.png *Source*: http://en.wikipedia.org/w/index.php?title=File:Poland_administrative_division_1975_literki.png *License*: unknown *Contributors*: w:en:User:Halibutt

File:Poland administrative division 1957 literki.PNG *Source*: http://en.wikipedia.org/w/index.php?title=File:Poland_administrative_division_1957_literki.PNG *License*: unknown *Contributors*: w:en:User:Halibutt

File:Poland administrative division 1922 literki.png *Source*: http://en.wikipedia.org/w/index.php?title=File:Poland_administrative_division_1922_literki.png *License*: unknown *Contributors*: w:en:User:Halibutt

File:KingdomofPoland1815.jpg *Source*: http://en.wikipedia.org/w/index.php?title=File:KingdomofPoland1815.jpg *License*: unknown *Contributors*: Mariusz Paździora

File:Rzeczpospolita voivodships.png *Source*: http://en.wikipedia.org/w/index.php?title=File:Rzeczpospolita_voivodships.png *License*: unknown *Contributors*: User:Halibutt

File:Irp.jpg *Source*: http://en.wikipedia.org/w/index.php?title=File:Irp.jpg *License*: unknown *Contributors*: User:Halibutt

Printed by Books on Demand GmbH, Norderstedt / Germany